XVI International Youth Scientific School on Actual Problems of Magnetic Resonance and its Applications 2013

AA002693

Journal of Physics: Conference Series Volume 478

Kazan, Russian Federation
21-25 October 2013

ISBN: 978-1-62993-940-7
ISSN: 1742-6588

Printed from e-media with permission by:

Curran Associates, Inc.
57 Morehouse Lane
Red Hook, NY 12571

Some format issues inherent in the e-media version may also appear in this print version.

Printed by Curran Associates, Inc. (2014)

For permission requests, please contact the Institute of Physics
at the address below.

Institute of Physics
Dirac House, Temple Back
Bristol BS1 6BE UK

Phone: 44 1 17 929 7481
Fax: 44 1 17 920 0979

techtracking@iop.org

Additional copies of this publication are available from:

Curran Associates, Inc.
57 Morehouse Lane
Red Hook, NY 12571 USA
Phone: 845-758-0400
Fax: 845-758-2634
Email: curran@proceedings.com
Web: www.proceedings.com

XVI International Youth Scientific School on Actual Problems of Magnetic Resonance and its Applications 2013

Journal of Physics: Conference Series Volume 478

Kazan, Russian Federation
21-25 October 2013

ISBN: 978-1-62993-940-7
ISSN: 1742-6588

IOP Institute of Physics

TABLE OF CONTENTS

PERSPECTIVE OF ZERO-FIELD ODMR TO STUDY NANO-BIOLOGICAL SYSTEMS 1
B. V. Yavkin, M. R. Gafurov, S. S. Kharintsev, G. V. Mamin, E. Goovaerts, M. Kh Salakhov, Yu N Osin, S. B. Orlinskii

INVESTIGATION OF ATHEROSCLEROTIC PLAQUE BY HIGH-FREQUENCY EPR 6
T. B. Biktagirov, Yu A. Chelyshev, M. R. Gafurov, G. V. Mamin, S. B. Orlinskii, Yu N. Osin, M. Kh Salakhov

HIGH-FREQUENCY EPR STUDY OF CRUDE OILS 11
M. A. Volodin, G. V. Mamin, V. V. Izotov, S. B. Orlinskii

MAGNON BOSE-EINSTEIN CONDENSATION AT INHOMOGENEOUS CONDITIONS 15
Yury Bunkov, Murat Tagirov

MOMENTS DISTRIBUTIONS OF SINGLE DYE MOLECULE SPECTRA IN A LOW-TEMPERATURE POLYMER: ANALYSIS OF SYSTEM ERGODICITY 24
T. A. Anikushina, A. V. Naumov

STATISTICS FOR FLUORESCENCE PHOTONS OF DONOR AND ACCEPTOR MOLECULES INVOLVED IN NON-RADIATIVE ENERGY TRANSFER (FRET) 29
A. L. Shchukina, I. S. Osad'Ko, I. Yu Eremchev

LUMINESCENCE PROPERTIES OF CDSE NANOCRYSTALLITES IN CADMIUM ALKANOATE GLASSES 39
K. A. Magarian, V. V. Fedyanin, K. R. Karimullin, I. A. Vasilieva, G. V. Klimusheva

CONTROL OF THE INDEX OF REFRACTION IN OPTICALLY DENSE MEDIUM 45
K. A. Barantsev, A. N. Litvinov

EFFECTIVE DEFECT MODE SUPPRESSION IN A MAGNETOPHOTONIC CRYSTALS IN THE MAGNETIC RESONANCE REGION 49
Svetlana V. Ellseeva, Vladimir A. Ostatochnikov, Dmitrij I. Sementsov

OPTICAL VIBRATIONS IN A CLOSED CHAIN OF COUPLED WAVEGUIDES OR CAVITIES 55
A. P. Sukhorukov, I. V. Savochkin

DERIVATION OF KINETIC EQUATIONS FROM NON-WIENER STOCHASTIC DIFFERENTIAL EQUATIONS 59
A. M. Basharov

BREATHER-LIKE PULSES IN A REDUCED MAXWELL-DUFFING MODEL 63
E. V. Kazantseva, A. I. Maimistov

METHODS OF CHARACTERIZATION OF SYNTHETIC OPAL FILMS 67
A. V. Koryukin, A. A. Akhmadeev, M. Kh Salakhov

SYNTHESIS OF NICKEL MICROWIRES WITH NANOSTRUCTURED SURFACE FOR ELECTRODES OF CHEMICAL CURRENT SOURCES 71
M. V. Morozov, A. Kh Gilmutdinov, M. Kh Salakhov

OPTICAL NEAR-FIELD SCATTERING IN A TIP-SAMPLE SYSTEM: A GREEN'S FUNCTION APPROACH 77
A. M. Rogov, A. R. Gazizov, S. S. Kharintsev, M. Kh Salakhov

DC-PULSED VOLTAGE ELECTROCHEMICAL METHOD BASED ON DUTY CYCLE SELF-CONTROL FOR PRODUCING TERS GOLD TIPS 82
V. E. Vasilchenko, S. S. Kharintsev, M. Kh Salakhov

ELECTRON REST MASS AND LINE SPECTRUM OF ATOMS IN THE PHOTONIC CRYSTAL MEDIUM 87
R. Kh Gainutdinov, M. A. Khamadeev, M. Kh Salakhov

NONLINEAR EFFECT IN PROCESSES OF THE INTERACTION OF ATOMS WITH A RESONANT LASER FIELD 92
R. Kh Gainutdinov, A. A. Mutygullina

UNSTABLE VACUUM AND SPECTRUM OF ATOMS WITH A SUPERHEAVY NUCLEUS 97
R. Kh Gainutdinov, A. A. Mutygullina, M. A. Khamadeev, A. S. Petrova

ELECTRON DYNAMICS AT GAAS-ALGAAS HETEROJUNCTION STUDIED BY ULTRAFAST SPECTROSCOPY 104
A. V. Leontyev, K. V. Ivanin, T. G. Mitrofanova, V. S. Lobkov, V. V. Samartsev

IMPLEMENTATION OF DOUBLE-PULSE LASER CONTROL IN OPTICAL KERR E 108
V. G. Nikiforov, A. G. Shmelev, G. M. Sa

HEAT-ASSISTED EFFECTS IN FERROMAGNETIC NANOPARTICLES 112
A. A. Bukharaev, D. A Bizyaev, T. F. Khanipov, N. I. Nurgazizov

FTIR INVESTIGATION OF MOLECULAR MOBILITY IN GLASSY EPOXY- AMINE OLIGOMERS WITH AZO-CHROMOPHORES 117
K. L. Nefedyeva, A. I. Fishman, M. Yu Balakina, T. A. Vakhonina
PUMP-PROBE STUDIES OF ABSORPTION SATURATION AND OPTICAL GAIN IN CE:LICAALF$_6$ ULTRAVIOLET ACTIVE MEDIUM 122
A. Galiev, V. Semashko, O. Akhtyamov, S. Shnaidman, M. Marisov
STUDYING THE FIDELITY OF QUANTUM MEMORY BASED ON CONTROL- FIELD ANGULAR SCANNING 126
L. Chicherova, A. Kalachev
MAGNETIC AND MAGNETOELASTIC PROPERTIES OF LIDYF$_4$ SINGLE CRYSTALS 132
I. V. Romanova, S. L. Korableva, V. I. Krotov, B. Z. Malkin, I R Mukhamedshinl, H. Suzuki, M. S. Tagirov
Author Index

Perspective of zero-field ODMR to study nano-biological systems

B V Yavkin[1], M R Gafurov[1], S S Kharintsev[1], G V Mamin[1], E Goovaerts[2], M Kh Salakhov[1], Yu N Osin[1], S B Orlinskii[1]

[1] Kazan Federal University, Kazan 420008, Russian Federation
[2] University of Antwerp, Antwerpen (Wilrijk) 2610, Belgium

E-mail: boris.yavkin@gmail.com

Abstract. This work presents an introduction, a short literature review as well as our recent optical and high field magnetic resonance experiments with regard to the applications of nitrogen paramagnetic defects in (nano)diamonds for biomedical related research. The perspectives of combination of optical and magnetic resonance (high field electron paramagnetic resonance) spectroscopic methods for the sensitive spatially resolved screening of electrical and magnetic gradients in biological tissues on the nanoscale level are discussed.

1. Introduction

Advancements in magnetic resonance (MR) have contributed immensely to a wide range of scientific areas from fundamental physics and chemistry to practical applications such as medical sciences, for example [1, 2]. Even though, MR methods have two inherent features that under some circumstances could be regarded as their disadvantages, e.g. low sensitivity and broad spectral lines. To overcome these obstacles, so called "brute-force" approach which utilizes low temperatures and extremely high magnetic fields is in use [3–5]. Alternative and complementing to the brute-force approach are the methods of double resonances. Though such popular methods as an electron-nuclear double resonance (ENDOR), electron-electron double resonance (DEER or ELDOR), and dynamic nuclear polarization (DNP) have proven their efficiency to study biological systems [6–8], none of the state of the art MR detection techniques combines high sensitivity, nanoscale spatial resolution and life sustaining working conditions, therefore limiting applications in life sciences. An illustration of the importance of nanoscale spatial resolution in the studies of biological substances – a part of the results of our micromorphological study of the aorta wall deposit from the male patient – is presented in figure 1 (see also an accompanying paper [9] for details). The corresponding microanalysis shows that the sample on the 10 μm scale is highly inhomogeneous. While section #1 and #4 of the investigated organomineralic matrix contain a detectable amount of Na and Cl, none of their traces were found in other sections. Nitrogen and iron in section #2 were revealed, while high calcification levels in sections #3 and #5 with the Ca content up to 6 wt. % were measured.

It is clear enough that for the research of inhomogeneous samples (biological or of any other kind) it is vital to have an experimental device with high spatial resolution as well as with high sensitivity. Let us have a look over the optically detected MR methods that could be capable to satisfy these two criteria.

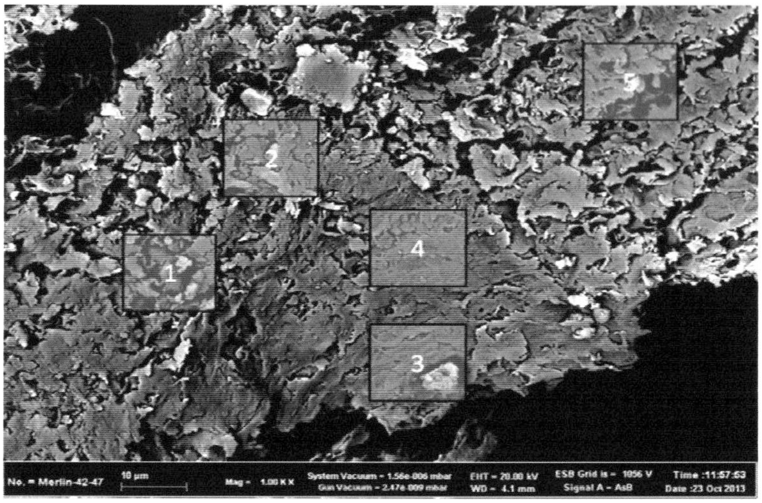

Figure 1. Electron microscopy of the part of the fatty deposit of the aorta wall of male patient gathered during the surgical operation. The squares mark the sections which were analyzed by elemental EDS.

2. ODMR: short history and basic concepts

ODMR (optically detected magnetic resonance) is a double resonance technique combining optical measurements (fluorescence, phosphorescence, absorption etc) with EPR. Bitter was the first who in 1949 theoretically showed that the resonance pumping of the EPR transitions should lead to the perceptible changes in polarization scheme of emission spectrum of mercury vapor [10]. Since that time the main area of the ODMR application is an investigation of different paramagnetic centers (often denoted as point defects) in semiconductors, layered structures and low-dimensional systems [11]. Schmidt and Van der Waals demonstrated that in some cases the ODMR signal could be detected in the vanishing magnetic fields if the paramagnetic system possesses zero-field splitting (ZFS) [12]. Resolution of the transition frequencies in zero-field ODMR is determined mainly by local environment of paramagnetic species, whereas EPR linewidths are much wider due to additional contribution of magnetic-field dependent interactions. A fruitful area of application of ZFS singlet-triplet states concerning biology has proven to be photosynthesis [13].

Due to the fact that the energies of optical quanta are much higher than the energies of microwaves quanta, ODMR is a quite sensitive technique compared to conventional EPR. For the same reason optical waves could be focused onto the much smaller spot on the sample. Therefore, ODMR studies could be done locally with high spatial resolution. Anisimov with his co-workers demonstrated that just 20 short-lived radical pairs could be detected by ODMR [14] while the amount of 10^9 spins are considered to be the lower detection limit for the commercial conventional EPR spectrometers. In 1993 two groups independently reported single spin detection on a single pentacene molecule in a p-terphenyl crystal [15, 16]. The experiments were done at T = 1.5 K - usual for the most ODMR experiments temperature to overcome the short relaxation times (lifetimes) of the centers studied.

Nobody could expect that the ODMR methods would be applied for the physiologically relevant investigations. Breakthrough was done in 1997 with the idea to use nitrogen-vancancy defects in diamonds for these purposes [17].

3. NV defects in (nano)diamonds

The nitrogen-vancancy (NV) center has recently become a high profile candidate for many different solid state applications, ranging from solid state quantum computing to magnetometry. NV centre in diamond consists of a nearest-neighbor pair of a nitrogen atom, which substitutes for a carbon atom and a lattice vacancy (figure 2). The center exists in two charge states: neutrally charged state NV^0 with electron spin S=1/2 and negatively charged state NV^- with electron spin S=1 (denoted as NV in this paper) [18–21].

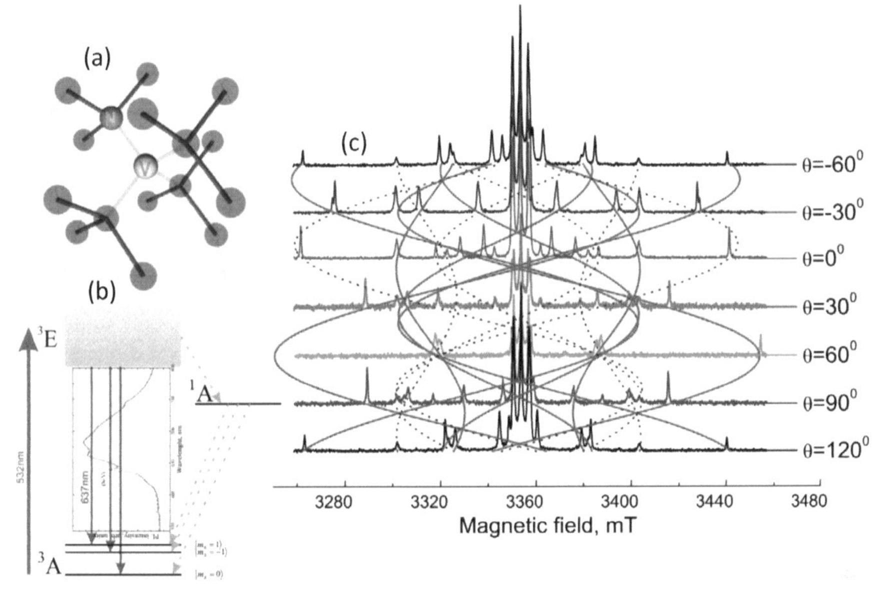

Figure 2. a) Schematic structure of NV center in diamond. b) Energy level structure and an example of the photoluminescence spectrum with ZPL lines shown at 637nm. (c) Example of the orientation dependence of 94 GHz electron-spin echo detected EPR spectra of N^0_S donor at RT (group of the central lines) and NV centers (highly anisotropic) in a detonation nanodiamond. Roadmap simulation for NV centers is shown by solid and dotted lines.

Among others, biological applications demand very bright and photostable fluorescence. The NV defects transform nanodiamonds into sensitive optical probes. The fluorescence spectrum of the NV is presented in figure 2. ZFS parameter of NV center is quite sensitive to the local magnetic [22] and electric fields [23] as well as to the temperature changes [24]. Moreover, since NV centers are atomic-sized point defects and can be localized in direct proximity to a diamond surface, they can be brought to within a few nanometers of samples, allowing for nanometric spatial resolution [25]. Nowadays, there are two main concepts of the zero-filed ODMR experiments at ambient conditions that exploit diamond thin film or functionalized nanodiamonds correspondingly.

First approach utilizes a diamond thin film with shallowly implanted nitrogen atoms converted afterwards to NV centers by using different procedures [26]. The lasts could be located very close to the film surface. That surface is covered by biological (or any other) material for investigation. The advantage of this method is high sensitivity because of the relatively high and controllable NV centers concentration. Disadvantage is a limited spatial resolution due to the finite minimal distance between the sensing NV center and area of interest in the material studied.

Second approach uses nanodiamond particles with NV centers inside. By the methods of chemical functionalization these nanoparticles could be directly bound to the desirable place in the biological tissue, so the distance between the NV center and measuring point could be small. This approach has a few disadvantages: decreased sensitivity due to the low NV concentration in nanodiamonds which depends on the nanoparticle size and functionalization procedure [27], and lower resolution because of physical motion of the nanoparticles in the biological tissues under physiological conditions [28].

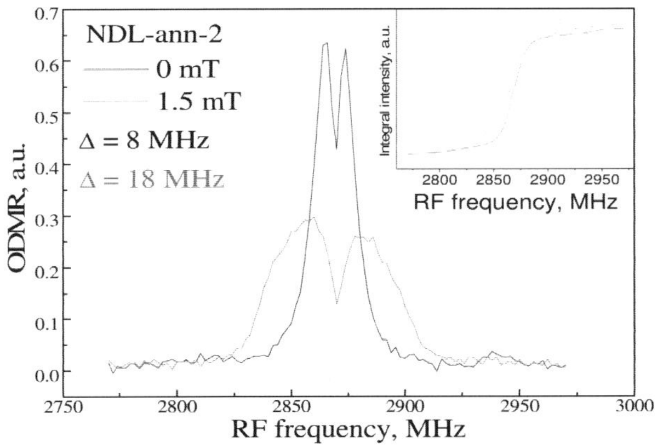

Figure 3. Low-frequency ODMR of NV centers in nanodiamonds, registered at RT in zero and low (1.5 mT) magnetic fields. Inset shows that integral intensity of lines in both cases are practically the same

Figure 3 presents the ODMR spectra of NV centers in the weak magnetic fields. The observed feature could be used for the mapping/screening of the distribution of the magnetic and electric fields caused by the inhomogeneities of the biological tissues. There are already a few promising attempts are done to combine the optical "single molecular" spectroscopy with ODMR [16] and their effectiveness is shown.

4. Materials and methods

For the details of EPR and microscopy experiments please have a look at the accompanying paper [9].

ODMR experiments in zero-field was done in continuous wave (CW) regime with simultaneous frequency sweep and modulation, using Rohde&Schwarz SMC 100A RF synthesizer as a source of microwave signal at frequencies of 2.87 GHz with maximal power of 0.5 W; laser illumination at 532 nm wavelength with power up to 150 mW in CW was provided by Nd:YAG laser from Wicked Lasers.

5. Conclusion

We are sure that ODMR approach combined with the modern techniques of the optical confocal microscopy would be effective for studying the wide range of biological objects at physiologically relevant conditions and would allow one to obtain new information about the objects and processes studied. Moreover, it should open a perspective to search for new materials which properties could be even more profitable for using them as nanosensors comparing to the NV-based sensors [29, 30].

References

[1] Burlaka A P, Ganusevich I I, Gafurov M R, Lukin S N and Sidorik E P 2013 *Cancer Microenviron.* (in press) DOI: 10.1007/s12307-013-0137-z
[2] Swartz H M et al 2004 *NMR Biomed.* **17** 335

[3] Yavkin B V et al 2012 *Phys. Chem. Chem. Phys.* **14** 2246
[4] Gafurov M R et al 2013 *Magn. Reson. Solids* **15** 13102
[5] Gafurov M, Lyubenova S, Denysenkov V, Ouari O, Karoui H, Le Moigne F, Tordo P and Prisner T 2010 *Appl. Magn. Reson.* **37** 505
[6] Orlinskii S B et al 2001 *Nanosci. Nanotechnol. Lett.* **3** 63
[7] Gafurov M R 2013 *Magn. Reson. Solids* **15** 13103
[8] Gafurov M, Denysenkov V, Prandolini M J and Prisner T 2012 *Appl Magn Reson* **43** 119
[9] Biktagirov T B, Chelyshev Yu A, Gafurov M R, Mamin G V, Orlinskii S B, Osin Yu N and Salakhov M Kh 2014 *J. Phys. Conf. Series* this issue 012002
[10] Bitter F 1949 *Phys. Rev.* **76** 833
[11] Spaeth J-M, Niklas J R and Bartram R H 1992 *Structural Analysis of Point Defects in Solids* (New York: Springer-Verlag) 367 p
[12] Schmidt J, Van der Waals J H 1968 *Chem Phys Lett* **2**(8) 640
[13] Carbonera D 2009 *Photosynthesis Res.* **102** 403
[14] Anisimov O A, Grigoryants V M, Molchanov V K, Molin Yu N 1979 *Chem Phys Lett* **66**(2) 265
[15] Kohler J, Disselhorst J A J M, Donckers M C J M, Groenen E J J, Schmidt J, Moerner W E 1993 *Nature* **363** 242
[16] Wrachtrup J, Borczyskowski C, Bernard J, Orrit M, Brown R 1993 *Nature* **363** 244
[17] Gruber A, Draebenstedt A, Tietz C, Fleury L, Wrachtrup J, Borczyskowski C 1997 *Science* **276** 2012
[18] Soltamova A A et al 2009 *Physica B: Condensed Matter* **404** (23) 4518
[19] Baranov P G et al 2009 *JETP letters* **89**(8) 409
[20] Baranov et al 2011 *SMALL* **7** 1533
[21] Yavkin B V et al 2013 *Appl. Magn. Reson.* **44** 1235
[22] Staudacher T, Shi F, Pezzagna S, Meijer J, Du J, Meriles C A, Reinhard F, Wrachtrup J 2013 *Science* **339** 561
[23] Dolde F et al 2011 *Nature Physics* **7** 459
[24] Kuscko G, Maurer P C, Yao N Y, Kubo M, Noh H J, Lo P K, Park H and Lukin M D 2013 *ArXiv*:1304.1068v1
[25] Balasubramanian G et al 2008 *Nature* **455** 648
[26] Acosta V M et al 2009 *Phys. Rev.* B **80** 115202
[27] Petrakova V et al 2012 *Adv. Funct. Mater.* **22**(4) 812
[28] Horowitz V R, Aleman B J, Cleland C A N, Awschalom D D 2012 *ArXiv*:1206.1573v1
[29] Kolesov R, Xia K, Reuter R, Jamali M, Stohr R, Inal T, Siyushev P and Wrachtrup J 2013 *Phys. Rev. Lett.* **111** 120502
[30] Haberle T, Schmid-Lorch D, Karrai K, Reinhard F and Wrachtrup J 2013 *Phys. Rev. Lett.* **111** 170801

Journal of Physics: Conference Series **478** (2013) 012002 doi:10.1088/1742-6596/478/1/012002

Investigation of atherosclerotic plaque by high-frequency EPR

T B Biktagirov [1], Yu A Chelyshev [1,2], M R Gafurov[1], G V Mamin [1], S B Orlinskii [1], Yu N Osin[1], M Kh Salakhov[1]

[1]Kazan Federal University, Kazan, 420008, Russia
[2]Kazan State Medical University, Kazan, 420012, Russia

E-mail: tbiktagirov@gmail.com

Abstract. We present a comparative study of samples of aorta walls from male patients with atherosclerosis and hydroxyapatite powders with the average size of crystallites of 30 nm synthesized by the wet precipitation technique by using 94 GHz pulsed EPR. Origin of the observed paramagnetic centers is discussed. Supported by the electron microscopy and microanalysis, it is shown that EPR spectra from the calcified biological tissues correlates with those obtained in inorganic hydroxyapatites. The hypothesis about the important role of (nano)hydroxyapatite in formation of the mineral deposits and atherosclerotic plaque instability is further sustained.

1. Introduction

Interest to this study is governed by the search for reliable prognostic markers for atherosclerotic plaque development. *Plaque* (also known as *atheroma*) stands for fatty deposits within the coronary arteries and other large arteries in arteriosclerotic cardiovascular diseases (ASVD). ASVD is a condition in which an artery within the heart wall thickens and hardens. According to the current understanding, a large number of diseases with totally different clinical signs are basically atherosclerosis related: myocardial infarction, stroke, abdominal aneurysms, lower limb ischemia, etc. The mortality rate in Russian Federation from the diseases of the blood circulatory system is reported to be about 800 on 100 000 (2009) [1-3].

Most of the acute manifestations of atherosclerosis share a common pathogenetic feature: rupture of an atherosclerotic plaque. Atherosclerotic plaque contains multiple components including lipidic, fibrotic, thrombotic, and calcific materials. The calcium deposits in cardiovascular tissues are comprised of various calcium salts, such as calcium carbonate, calcium oxalate, calcium phosphate, calcium pyrophosphate, and hydroxyapatite (HAp). It was proposed that the microcalcification is tightly connected with the instability of the plaque, leading to the plaque rupture [4]. Namely, as it was assumed in [5], it goes hand-to-hand with the presence and size of the HAp component in the organomineralic matrix of aorta deposits and, therefore, the nanosized HAp components of the aorta tissues could serve at least as markers for the early diagnosis of the plaque rupture risks.

To support or kick out the last hypothesis, in present work we use the advantages of the increased resolution and sensitivity of high-frequency (94 GHz) electron paramagnetic resonance (EPR) together with energy dispersive spectrometer (EDS) analysis combined with the scanning electron microscopy (SEM) to compare the results from the samples of aorta walls from male patients with atherosclerosis at different stages with those from the modified synthetic HAp.

2. Materials and Methods

Plaque tissues of aorta walls from male patients with atherosclerosis at different stages were gathered in Interregional Clinic and Diagnostic Centre, Kazan, Russia following the rules and under control of the responsible Ethics Committee. Sample preparation was the same as described in paper [5]. The HAp-based powders with the average size of crystallites of 30 nm were synthesized in the group of V.I. Putlyaev (Department of Material Sciences, Moscow State University, Moscow, Russia) by the wet (precipitation) chemical procedure involving the aqueous solutions of calcium nitrate, diammonium hydrogen phosphate and ammonium hydroxide:

$$10\,Ca(NO_3)_2 + 6\,(NH_4)_2HPO_4 + 8\,NH_4OH \rightarrow Ca_{10}(PO_4)_6(OH)_2\downarrow + 20\,NH_4NO_3 + 6\,H_2O. \qquad (1)$$

X-ray diffraction (XRD) spectra show that all the synthesized samples contain only one phase corresponding to the space group $P6_3/m$ with the parameters of the unit cell close to $a \approx b \approx 0.942$ nm and $c = 0.688$ nm typical for the bulk crystals of the hydroxyapatite [6]. Corresponding EDS investigations ensure the ratio Ca/P = 1.66(2) within the accuracy of our measurements in the powder grains serving as an evidence of the stoichiometry of the investigated HAp species. Details of the synthesis of the HAp samples and their modifications containing manganese and carbonate ions, post-synthesis treatment and results of the analytical characterization are described in [7-9].

The micromorphology of the biological samples, their Ca- and P- contents was examined and defined by scanning electron microscopy (SEM, Merlin, Carl Zeiss) combined with the electron dispersive spectrometer (EDS, AZtec, Oxford Instruments) in the group of Yu.N. Osin (Interdisciplinary Center of Analytical Microscopy, Kazan Federal University, Kazan, Russia). The samples were coated by Au/Pd = 80:20 thin film of about 20 nm, (registration conditions: 20 kV, 300 pA, imaging mode: detectors SE, Inlens, AsB, EDS).

To create stable paramagnetic centers in EPR silent samples, X-ray irradiation of the species was performed using URS-55 tube (U = 55kV, I = 16mA, W-anticathode) at room temperature with the estimated dose of 10 kGy to have an ability to apply different EPR techniques. XRD patterns before and after X-ray irradiation were checked in Bruker D2 Phaser diffractometer (Bruker, Germany) and were the same.

Pulsed EPR done in high-frequency (94 GHz, W-band) Bruker Elexsys 680 spectrometer (Bruker, Germany). EPR spectra were recorded by means of field-swept two-pulse echo sequence $\pi/2$-τ-π with the minimal pulse length of π pulse of 4 ns and time delay $\tau = 220$ ns.

3. Results and Discussion

Strong correlation between the calcification of the aorta samples and possibility to obtain EPR was revealed in our experiments. EPR spectra in solid samples at room temperature were obtained only in the calcified tissues. Figure 1 presents the micromorphology of the sample that gives the detectable EPR signals before as well as after X-ray irradiation. Some segments were chosen for the EDS microanalysis (marked by squares in figure 1). C, O, Ca and P were detected there. While the content of Ca reached up to 25 wt %, the ratio Ca/P (in at %) for them was found to be in the range (1.80-2.20). The last values are slightly higher than those for the stoichiometric hydroxyapatite (Ca/P = 1.66) and are within the range commonly reported for the biogenic carbonated hydroxyapatite (1.70-2.30; see [10] and references therein). Thus one could expect the obtained values to suggest that (10-20) % of PO_4 groups in HAp are substituted by the carbonate, even though the presence of the other forms of calcium phosphates in the studied samples also should not be excluded. In our opinion, this finding could serve as a piece of evidence that the investigated aorta tissues contain (hydroxy)apatite like material. Therefore, we have supposed that the EPR signals are due to some paramagnetic centers incorporated into the HAp. (An example of SEM and EDS for the sample from the aorta wall that gives a low-intensive EPR signals is presented in accompanying paper [11]).

Different paramagnetic centers could be detected in the synthetic and natural hydroxyapatite based materials but only few of them in the HAp of biogenic origin, especially without X-ray irradiation. Figure 2 shows the electron spin echo (ESE) signal detected during the magnetic field sweep (field-sweep ESE) in W-band at room temperature of atherosclerotic plaque and synthesized

$Mn_{0.0005}Ca_{9.9995}(PO_4)_6(OH)_2$ (MnHAp) powder with the average grain size of 30 nm. The obtained six-line pattern signal is undoubtedly stemmed from the hyperfine interaction between electron (S = 5/2) and nuclear (I = 5/2) spins of the incorporated manganese (Mn^{2+}) in non-oriented powders. Taking into account the accuracy of our experiments originating from the different sources of the magnetic field inhomogeneity at the sites of the studied paramagnetic centers, the closeness of the observed spectroscopic parameters (g-factor of 2.001(1) and hyperfine splitting constant of 9.3(3) mT) points out that manganese in aorta samples is most probably located in HAp component of the plaque organomineralic matrix.

Figure 1. Electron microscopy of the part of the calcified deposit from the aorta wall. The squares mark the segments chosen for the microanalysis.

The small concentration of manganese neither in biological tissues nor in synthetic HAp was observed by the microanalysis or by the conventional X-band (9 GHz frequency) EPR. To detect the trace amounts of elements other methods of atomic spectroscopy (like AAS – atomic absorption spectroscopy) often linked with the mass-spectrometry are usually applied [12-14] which are initially the destructive techniques. In this sense, the presented results emphasize the advantages of high-frequency EPR as a sensitive, requiring very low (of less than 1 μg) amount of tissue, non-destructive tool to study different paramagnetic species in different materials, their (electronic) structure as well as a variety of types of interactions between them [15, 16].

After X-ray irradiation, in contrary, a variety of radiation-induced defects could be created which EPR spectra might overlap each other complicating, therefore, their interpretation [6, 17]. In this context, by choosing the appropriate parameters of the detection in the pulsed mode (such as a temperature, pulse lengths, etc.), one can distinguish between the paramagnetic species with different relaxation times. In such a manner we could manage to separate the carbonate spectra which are presented in figure 3. The spectra are due to the most stable and long-relaxing CO_2^- radical [5, 17]. As a reference, the carbonated HAp with the carbonation level of about 13(1) wt % was taken.

The relationship between a few percent carbonate and its location in the lattice structure of apatite-like materials is still a "bone of contention" both from the theoretical and experimental points of view [7, 17], while the location of manganese ions is believed to be determined perfectly [7] . We hope that our further experiments could give a key to experimental determination of carbonate location by advanced EPR methods [5, 7, 16, 18].

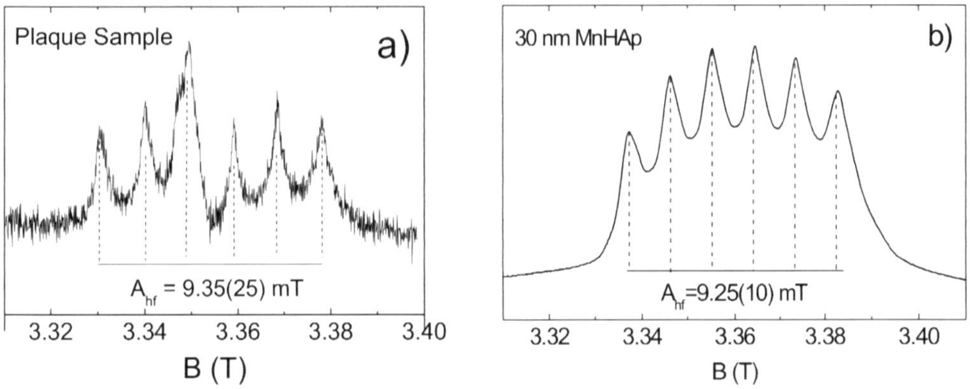

Figure 2. ESE-detected EPR spectra of the atherosclerotic plaque (a) and of the nano-MnHAp (b) at T = 293 K.

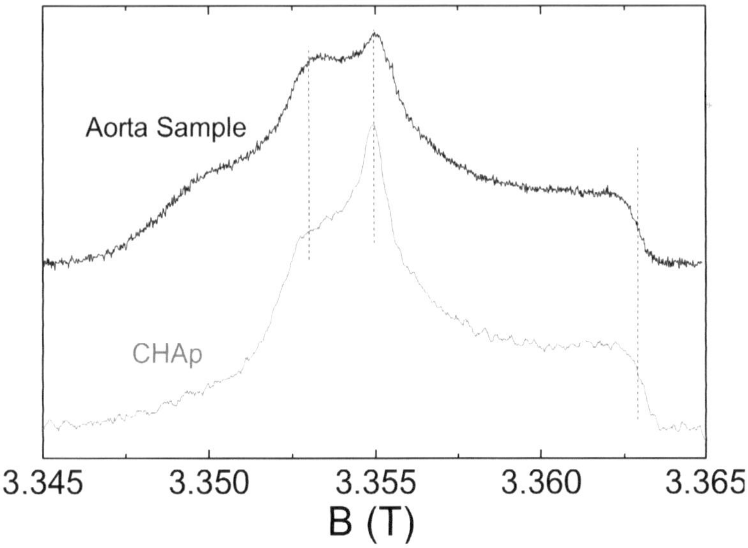

Figure 3. ESE-detected EPR spectra of the long-relaxing CO_2^- radiation-induced radical in aorta sample (upper curve) and in the carbonated HAp (CHAp, lower curve) at T = 293 K.

4. Summary

1. A high sensitivity of modern commercially available high-frequency spectrometers along with the pulsed methods open new horizons for using EPR methods in biomedical research. In particular, for early ASVD diagnosis.
2. EPR of manganese ions (Mn^{2+}) in the atherosclerotic plaques was detected. It is shown that the

spectroscopic parameters of Mn^{2+} ions are practically the same for the investigated atherosclerotic plaques and synthesized HAp nanoparticels.
3. EPR of the radiation induced CO_2^- radical in the atherosclerotic plaque was separated. The spectroscopic parameters of the radical are revealed to be the same as in the carbonated HAp.
4. A correlation between the calcification of the aorta wall deposits and intensity (possibility to detect) of the mentioned above EPR signals is observed.

References

[1] Petrukhin I S and Lunina E Y 2012 *Public Health Reviews* **33** 436
[2] Weissberg P L 2000 *Heart* **83** 247
[3] World Health Organization 2011 *Global Atlas on cardiovascular disease prevention and control* (Geneva) 155 p
[4] Vengrenyuk Y, Carlier S, Xanthos S, Cardoso L, Ganatos P, Virmani R, Einav Sh, Gilchrist L and Weinbaum S 2006 *PNAS* **103** 14678
[5] Abdul'yanov V A, Galiullina L F, Galyavich A S, Izotov V G, Mamin G V, Orlinskii S B, Rodionov A A, Salakhov M Kh, Silkin N I, Sitdikova L M, Khairullin R N and Chelyshev Yu A 2008 *JETP Lett.* **88** 69
[6] Elliott J C 1994 *Studies in Inorganic Chemistry* **18** 1
[7] Yavkin B V, Mamin G V, Orlinskii S B, Gafurov M R, Salakhov M Kh, Biktagirov T B, Klimashina E S, Putlayev V I, Tretyakov Yu D and Silkin N I 2012 *Phys. Chem. Chem. Phys.* **14** 2246
[8] Gafurov M R, Yavkin B V, Biktagirov T B, Mamin G V, Orlinskii S B, Izotov V V, Salakhov M Kh, Klimashina E S, Putlayev V I, Abdulyanov V A, Ignatjev I M, Khairullin R N, Zamochkin A V and Chelyshev Yu A 2013 *Magn. Reson. Solids* **15** 13102
[9] Kovaleva E S, Shabanov M P, Putlayev V I, Filippov Y Y, Tretyakov Y D and Ivanov V K 2008 *Mat-wiss u Werkstofftech* **39** 822-829
[10] Seah R K, Garland M, Loo J S and Widjaja E 2009 *Analytical chemistry* **81** 1442-1449.
[11] Yavkin B V, Gafurov M R, Kharintsev S S, Mamin G V, Goovaerts E, Salakhov M Kh, Osin Yu N, Orlinskii S B 2014. Perspective of zero-field ODMR to study nano-biological system 2014 *J Phys Conf Series* this issue 012001
[12] Hermann G, Trenin A, Matz R, Gafurov M, Gilmutdinov A K, Nagulin K Y, Frech W, Björn E, Grinshtein I and Vasilieva L 2004 *Spectrochimica Acta - Part B Atomic Spectroscopy* **59** 737-748
[13] Trenin A, Gafurov M, Gilmutdinov A Kh and Hermann G 2007 *Spectrochimica Acta Part B: Atomic Spectroscopy* **62** 231-241
[14] Lozhkin A P, Biktagirov T B, Abdul'yanov V A, Voloshin A V, Silkin N I, Khairullin R N, Salakhov M Kh and Ilinskaya O N 2010 *Doklady Biochemistry and Biophysics* **434** 254-256
[15] Orlinskii S B, Bogomolov R S, Kiyamova A M, Yavkin B V, Mamin G M, Turner S, van Tendeloo G, Shiryaev A A, Vlasov I I and Shenderova O 2011 *Nanoscience and Nanotechnology Letters* **3** 63-67
[16] Gafurov M, Lyubenova S, Denysenkov V, Ouari O, Karoui H, Le Moigne F, Tordo P and Prisner T 2010 *Appl Magn Reson* **37** 505-514
[17] Fattibene P and Callens F 2010 *Appl. Radiat. Isotopes* **68** 2033–2116
[18] Gafurov M, Denysenkov V, Prandolini M J and Prisner T F 2012 *Appl Magn Reson* **43** 119-128

Journal of Physics: Conference Series **478** (2013) 012003 doi:10.1088/1742-6596/478/1/012003

High-frequency EPR study of crude oils

M A Volodin, G V Mamin, V V Izotov and S B Orlinskii

Kazan Federal University, Kremlevskaya 18, 420008 Kazan, Russia

E-mail: volodinmikhail@yandex.ru

Abstract. Four different samples of crude oil were studied by means of high-frequency W-band (94 GHz) electron paramagnetic resonance (EPR) spectroscopy with the aim to develop new methods of crude oil quality control. High spectral resolution of W-band allowed to avoid an overlap of spectra contributors. The ratio K between the integral intensity of the low-field EPR component of the vanadyl complexes to that of free radical line was chosen as an attribute of each sample. Using the K-parameters and EPR spectra simulations the crude oil leaking between adjacent horizons is shown. Pulsed EPR experiments allowed detecting free radicals signals only. It is demonstrated that the extracted transverse relaxation time could be used as an additional parameter which characterizes the origin of the crude oil and nature of the oil paramagnetic centers.

1. Introduction

Due to the great heterogeniety of properties and composition of crude oils from different oilfields, a task to develop the methods of monitoring the characteristics of crude oils is still of a great importance. Electron paramagnetic resonance (EPR) is known as one of the informative tools for oil research but the vast majority of EPR applications were and are restricted to the conventional X-band (9 GHz) measurements. The main scope of the present work is to demonstrate some capabilities of the modern high-frequency (W-band, 94 GHz) EPR spectroscopy for investigation of crude oils.

2. Materials and Methods

A list of the investigated samples is presented in table 1. Samples 1 and 2 are from the same oilfield and well but extracted from different depths. Sample 4 has been investigated to prove the suggestion about the oil leaking from Vereian horizon to Bashkirian horizon.

Table 1. Studied samples.

Sample No.	Oilfield	Well No.	Horizon
1	Akanskoe	2023	Bashkirian (bottom)
2	Akanskoe	2023	Bashkirian (top)
3	Yamashinskoe	7220	Tournai
4	Akanskoe	27	Vereian

All samples were studied at room temperature both in the continuous wave (CW) and pulsed modes of the commercial Bruker Elexsys-680 spectrometer. The measurements were performed in the microwave W-band (94 GHz). In the CW mode the spectra were obtained utilizing a standard approach with a double modulation of an applied magnetic field B_0 [1, 2].

3. Results

In figure. 1 the CW EPR spectrum of sample 3 is presented which is a typical one for the whole series of samples studied. EPR spectra of oil are due to the hyperfine structure of vanadyl complex VO^{2+} ($^{51}V^{4+}$, $3d^1$, $S = 1/2$, $I = 7/2$, $g_{||} = 1.9640 \pm 0.0005$, $g_{\perp} = 1.9840 \pm 0.0005$) and from the single line of carbon free radicals (FR), which is in agreement with results of [3, 4, 5]. Atoms in the vanadyl-porphyrins are arranged practically in a plane defining thus the g-factor and hyperfine A tensor have an axial symmetry [4, 5, 6]. The EPR spectrum of the vanadyl-ions consists of the 16 lines representing the 2×8 hyperfine patterns for the parallel and the perpendicular complex orientations.

An important consequence of the microwave frequency increase revealed in the observed CW EPR spectra in comparison to those measured in the X-band is an almost total spectral resolution between the signals of the vanadyl-ions and the FRs: even a small difference in the g-factor values is enough to avoid the overlap of the two main contributors to the spectra.

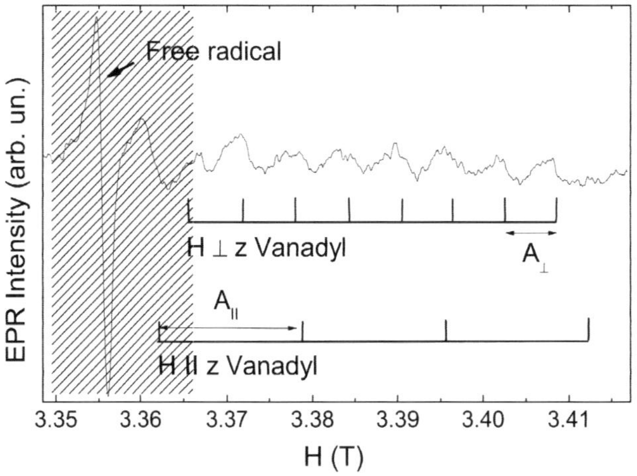

Figure 1. The CW EPR spectrum of sample 3. Other four lines of vanadyl in parallel orientation are outside of scale.

For comparison of CW EPR spectra from different oil samples we have considered the ratio K of vanadyl line integral intensity to FR line integral intensity. The difference in the spectra intensity is clearly visible in figure. 2 and figure. 3 (shaded area of figure. 1). table 2 contains calculated values of K-parameter for all studied oil samples. EasySpin subroutine module was used for EPR spectra simulations [7].

Samples 1 and 2 extracted from the different depth of the same well have different values of K parameter that could indicate the destruction of vanadyl-porphyrins in the oxidation processes. Indeed, the rate of the oxidation processes depends on the occurrence depth and, as a rule, increases towards to the earth surface [8]. As followed from the signal intensities, the concentration of paramagnetic species in sample 2 is significantly higher than that in sample 1. The presumable oil leaking from Vereian horizon to the top of Bashkirian horizon can be confirmed by the same lineshapes and values of K-parameter for samples 2 and 4.

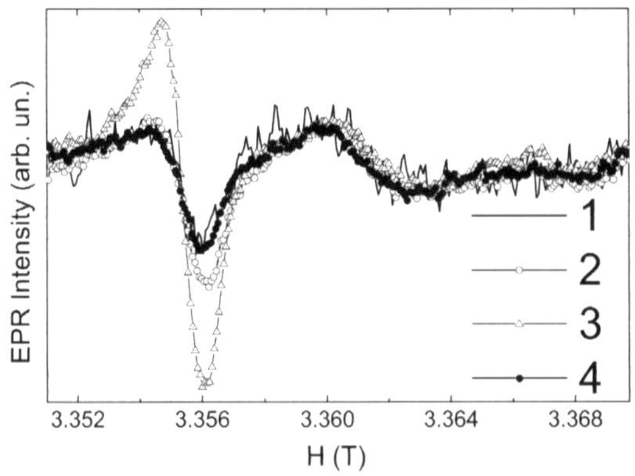

Figure 2. The CW EPR spectra of FR and low-field vanadyl component for all samples.

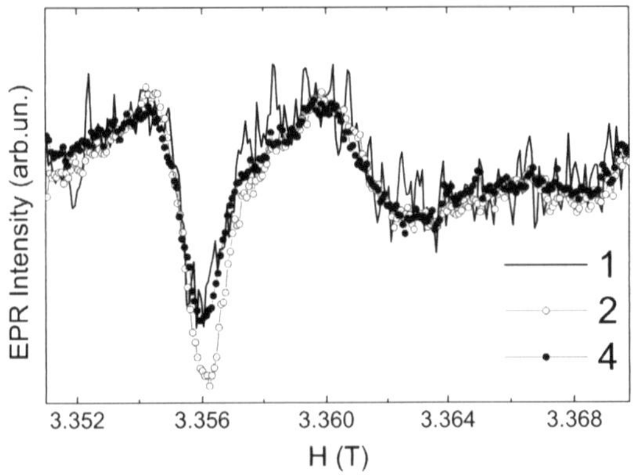

Figure 3. The CW EPR spectra of FR and low-field vanadyl component for samples 1, 2, 4.

Table 2. K-parameter values.

Sample No.	1	2	3	4
K-parameter	39.2	16.3	8.6	20.2

For getting the detailed information about the FR, electronic spin-echo (ESE) detected EPR spectra of FR were measured [9, 10]. ESE-detected EPR spectrum for oil sample 2 is shown in figure. 4.

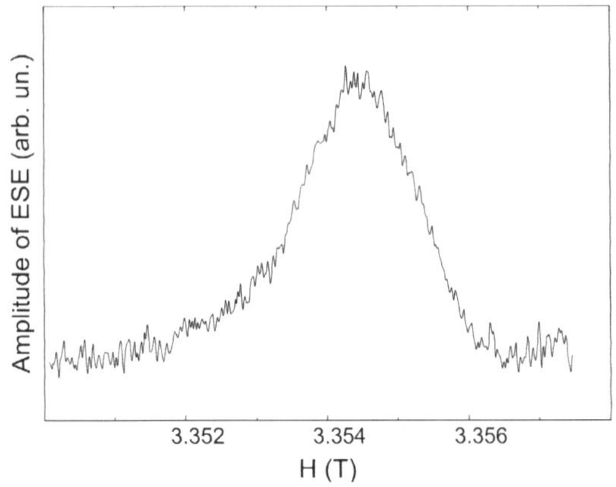

Figure 4. ESE-detected EPR spectrum of sample 2 ($g_{||} = 2.0028 \pm 0.0005$, $g_{\perp} = 2.0040 \pm 0.0005$).

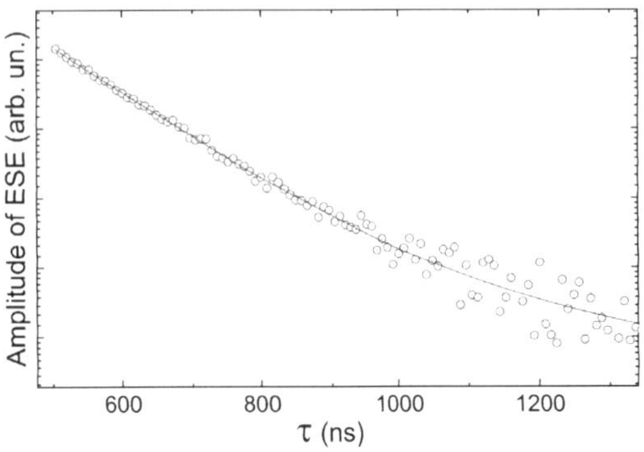

Figure 5. ESE decay for sample 3.

In ESE-detected spectra an inhomogeneous single line caused by presence of FR only is registered. Due to their short relaxation times [3] signals from vanadyl-ions were not obtained. The ESE-signal for sample 1 had not been observed probably due to low concentration of paramagnetic species.

Transverse relaxation was studied tracking the primary ESE amplitude while varying τ [9, 10]. The observed transverse magnetization decay for sample 3 is shown in figure. 5. Values of transverse relaxation times for all samples are represented in table 3.

Table 3. Transverse relaxation times of FR.

Sample No.	1	2	3	4
T_2, ns	-	185 ± 3	154 ± 2	197 ± 5

The possibility to detect FR ESE in liquid samples of crude oil and its inhomogeneous broadening indicates that FR is located within the large molecular clusters of asphaltenes. Otherwise transverse relaxation would be significantly shorter and unreachable for measuring by our equipment at RT due to the averaging caused by high mobility of paramagnetic species. Slight difference in the values of transverse relaxation time (table 3) between different samples can be explained by differences in sample viscosities.

4. Conclusion

Application of the high-frequency (W-band) EPR allows to resolve spectrally the free radicals and the vanadyl-ions components of the EPR spectra of crude oil. It gives an opportunity to gain a deeper insight to the origin of the paramagnetic centers important both for the industrial applications and fundamental research.

The ratio of integral intensities of vanadyl line to FR line (K-parameter) and transverse relaxation time of FR allow to distinguish oil types and could serve as fingerprints of the hydrocarbon origin.

Acknowledgments

The authors are grateful to Dr. M. R. Gafurov for fruitful discussions and careful reading of the manuscript.

The work was performed with a partial financial support from the Russian Ministry of Education and Science (project No. 02.G25.31.0029).

References

[1] Poole C P 1967 *Electron Paramagnetic Resonance. Comprehensive Treatise on Experimental Techniques* (John Wiley & Sons)
[2] Weil J A and Bolton J R *Electron Paramagnetic Resonance: Elementary Theory and Practical Applications* 2nd ed
[3] Volodin M A, Gafurov M R, Mamin G V, Orlinskii S B, Murzakaev M V and Yusupova T N 2013 *Oil Industry* **6** 44–47
[4] Gilinskaya L G 2008 *Journal of Structural Chemistry* **49** 245–254
[5] Dechaine G P and Gray M R 2010 *Energy Fuels* **24** 2795–2808
[6] Espinosa M, Campero A and Salcedo R 2001 *Inorg. Chem.* **40** 4543–4549
[7] Stoll S and Schweiger A 2006 *J. Magn. Reson.* **178** 42–55
[8] Russell W L *Principles of Petroleum Geology* 2nd ed
[9] Salikhov K M, Semenov A G and Tsvetkov Y D 1976 *Electron Spin Echo and Its Applications* (Novosibirsk: Nauka)
[10] Schweiger A and Jeschke G 2001 *Principles of Pulse Electron Paramagnetic Resonance* (Oxford University Press)

Magnon Bose-Einstein condensation at inhomogeneous conditions

Yury Bunkov[1,2] **and Murat Tagirov**[2]

1) Institut Néel, CNRS et Université Joseph Fourier, F-38042 Grenoble, France
2) Kazan (Volga region) Federal University, 420008, Kremlyovskaja 18, Kazan, Russia

E-mail: yuriy.bunkov@grenoble.cnrs.fr

Abstract. The Spin Supercurrent and Bose-Einstein condensation of magnons, similar to an atomic BEC, was discovered in superfluid ^{3}He-B, which is characterized by absolute purity. Later this phenomena were observed in a few magnetically ordered materials with different types of impurities. In this article we will review the properties of magnon BEC in a presence of impurities and defects.

1. Introduction

The conventional magnon BEC, the phase-coherent precession of magnetization was discovered for the first time in superfluid ^{3}He-B in 1984 [1]. It manifests itself by a region, where the magnetization is deflected on a large angle (about 104°) and is precessing coherently even in an inhomogeneous magnetic field. This state was named Homogeneously Precessing Domain (HPD). The transverse component of magnetization in HPD is described by the wave function $S_\perp e^{i\omega t+\phi}$. It possesses all the properties of the spin superfluidity. The spatial gradient of phase ϕ leads to a spin supercurrent which transports the magnetization. There were observed: phase-slip processes at the critical current [2]; spin current Josephson effect [3]; spin current vortex [4]; Goldstone modes [5, 6, 7], etc. The long standing quest of nontrivial magnetic relaxation in superfluid ^{3}He have been resolved [8]. The comprehensive review of spin superfluidity in ^{3}He-B one can find in [9] and recent one in [10, 11, 12].

There is another phenomenon named in some publications as magnon BEC (see [13]). In reality it is the special case of magnetic phase transitions which could be described in terms of magnon's equilibrium density. Near the phase transition, the magnon density increases and, at a certain temperature, a new magnetic state appears. Owing the Bose statistics of magnons, this phenomenon is described via similar mathematics as that for the conventional atomic BEC. However, in fact, there exists the principal difference. In the case of a conventional BEC, we deal with the state of excitations (atoms) in a quantum vacuum (Universe). The BEC state of atoms does not change the vacuum background. Similarly, in the case of magnon BEC we are describing here, the ground magnetic states do not change. What is being condensed is the non-equilibrium magnons which were created by NMR. The conventional BEC is known to be forbidden for the equilibrium gas of excitations (see [14]). The use of the term "magnon BEC" for magnetic phase transitions, described in review [13], is thus misleading.

There are many new physical phenomena related to the Bose condensation of magnons, which have been observed after the discovery of HPD. For the last 25 years there was found 5 different magnon BEC states in superfluid ^{3}He [15]. These include in particular compact objects – coherently precessing states trapped by orbital texture [16]. At small number N of the pumped magnons, the system is similar to the Bose condensate of the ultracold atoms in harmonic traps, while at larger N the analog of the Q-ball in particle physics develops [17].

A Q-ball is a non-topological soliton solution in field theories containing a complex scalar field Ψ. Q-balls are stabilized due to the conservation of the global $U(1)$ charge Q [18]. They are formed due to suitable attractive interaction that binds the quanta of Ψ-field into a large compact object. In some modern SUSY scenarios Q-balls are considered as a heavy particle-like objects, with Q being the baryon and/or lepton number. For many conceivable alternatives, Q-balls may contribute significantly to the dark matter and baryon contents of the Universe, as described in review [19]. Stable cosmological Q-balls can be searched for in existing and planned experiments [20].

The Q-ball is a rather general physical object, which in principle can be formed in condensed matter systems. In particular, Q-balls were suggested in the atomic Bose-Einstein condensates [21]. In ^{3}He-B, the Q-balls are formed as special states of phase coherent precession of magnetization. The role of the Q-charge is played by the projection $\mathcal{S}_z$ of the total spin of the system on the axis of magnetic field, which is a rather well conserved quantity at low temperature, or which is the same the magnon number N. At the quantum level, this Q-ball is a compact object formed by magnons – quanta of the corresponding Ψ-field.

In ^{3}He-B the Q-balls are formed at low temperatures, when homogeneous magnon BEC in the form of Homogeneously Precessing Domain (HPD) becomes unstable due to parametric Suhl instability [22, 23, 24]. At low temperatures the condensate can be formed only in a trap, similar to that in atomic gases [25], and the Q-balls are either formed in these traps or dig their own trap.

The Q-ball in superfluid ^{3}He-B was discovered by case as a very strange coherent signal of small amplitude (below 10% of HPD) but extremely long (up to minute) [26]. Later it was found that the frequency of the signal grows up, not down, as in the case of HPD [27]. It means that the mechanism of Q-ball formation is very different than HPD. In the works [28, 29] the steady state of Q-balls have been maintained by CW RF pumping. Again, in contrast with HPD, the signal was excited by sweep field up (frequency down). Finally the signal was explained as a formation of a Q-ball in a texture trap [17]. It was found that the Q-ball can be excited even by an off-resonance excitation [16, 30]. The recent detailed experimental investigations of Q-balls formed in the specially prepared and the well controlled traps were made in [31].

2. BEC of ^{3}He in aerogel

The further progress of the investigations of BEC states in superfluid ^{3}He is related to some kind of artificial impurities. This became possible by immersing the superfluid ^{3}He in a very porous material called aerogel. HPD spectroscopy proved to be extremely useful for the investigation of the superfluid order parameter in this novel system – superfluid ^{3}He confined in aerogel [32, 33, 34, 35, 36]. By squeezing or stretching the aerogel sample, one creates the global anisotropy which captures the orbital vector $\hat{\mathbf{l}}$. This allows to orient the orbital vector $\hat{\mathbf{l}}$ in the desirable direction with respect to magnetic field [37, 38].

In this conditions we are able experimentally reorient the orbital momentum of superfluid ^{3}He-B perpendicular to a magnetic field. For the transverse orientation of $\hat{\mathbf{l}}$ two new magnon states have been identified. One of them exists at $|\Psi|^2 < S/\hbar$ and has the following form of spin-orbit interaction obtained (we omit for simplicity the constant term):

$$F_{\rm so}\left(\Psi\right)_{l=0} = -\frac{\chi}{4\gamma^2}\Omega_L^2\left(\frac{|\Psi|^2}{S}-\frac{4}{5}\right)^2 , \ |\Psi|^2 < \frac{S}{\hbar}, \tag{1}$$

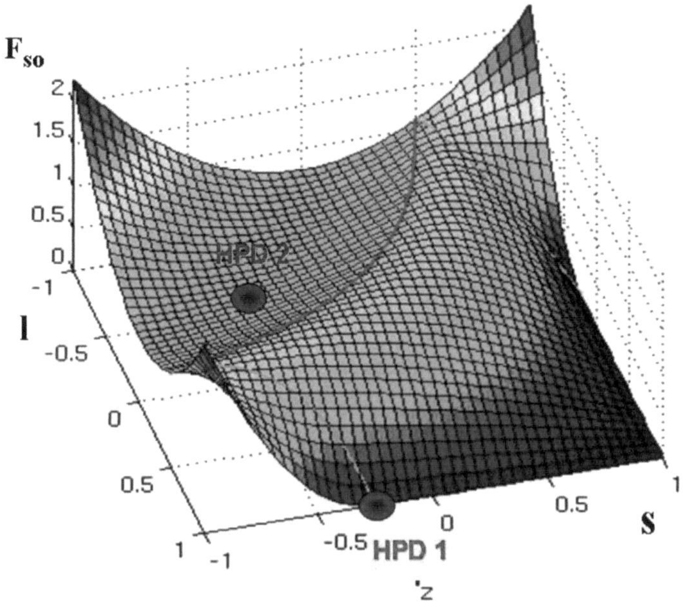

Figure 1. The profile of the spin-orbit energy in ^{3}He-B as a function of $S = \cos\beta$ and orbital variable l, where β is the tipping angle of precession and l is the projection of the orbital angular momentum of a Cooper pair on the direction of magnetic field. Spontaneous phase-coherent precession emerging at $l = 1$ and $S \approx -1/4$ is called Homegeneously Precessing Domain (HPD). The second BEC state (HPD2) is emerging at $l = 0$ and $S < 0$.

where χ is the susceptibility and Ω_L - Legget frequency, the frequency of the longitudinal mode of magnetic resonance in superfluid ^{3}He.

This state has an attractive interaction between magnons, and is unstable since the compressibility β_M of the magnon gas is negative: $d^2\epsilon/dn^2 < 0$.

The other state exists at $|\Psi|^2 > S/\hbar$ and has the following form of spin-orbit interaction

$$F_{\rm so}\left(\Psi\right)_{l=0} = \frac{\chi}{20\gamma^2}\Omega_L^2\left(\frac{|\Psi|^2}{S} - 2\right)^2 \quad , \quad |\Psi|^2 > \frac{S}{\hbar} \ . \tag{2}$$

This state has repulsive interaction between magnons and is stable. The magnon BEC formation under these conditions has been observed [39].

As in the case of ^{3}He-B, all the information on the ^{3}He-A order parameter needed to study the coherent precession is encoded in the spin-orbit interaction. For ^{3}He-A, the spin-orbit interaction averaged over the fast precession has the following form [15]:

$$F_{\rm so}\left(|\Psi|\right) = \frac{\chi\Omega_L^2}{4\gamma^2} \times$$
$$\left[-2\frac{|\Psi|^2}{S} + \frac{|\Psi|^4}{S^2} + \left(-2 + 4\frac{|\Psi|^2}{S} - \frac{7}{4}\frac{|\Psi|^4}{S^2}\right)(1 - l^2)\right] \tag{3}$$

In a static bulk ^{3}He-A, when $\Psi = 0$, the spin-orbit energy $F_{\rm so}$ in Eq.(3) is minimized when the orbital vector $\hat{\mathbf{l}}$ is perpendicular to magnetic field, i.e. for $l = 0$. Then one has

$$F_{\rm so}\left(|\Psi|, l=0\right) = \frac{\chi\Omega_L^2}{4\gamma^2}\left[-2 + 2\frac{|\Psi|^2}{S} - \frac{3}{4}\frac{|\Psi|^4}{S^2}\right], \qquad (4)$$

with a negative quartic term. The attractive interaction between magnons destabilizes the BEC, which means that homogeneous precession of magnetization in ^{3}He-A becomes unstable. This instability predicted by Fomin [40] was experimentally confirmed in Ref. [41].

However, as follows from (3), at sufficiently large magnon density $n = |\Psi|^2$

$$\frac{8+\sqrt{8}}{7}S > n > \frac{8-\sqrt{8}}{7}S , \qquad (5)$$

the factor in front of l^2 becomes negative. Therefore it becomes energetically favorable to orient the orbital momentum $\hat{\mathbf{l}}$ along the magnetic field ($l = 1$). For this orientation one obtains the Ginzburg-Landau free energy with

$$F_{\rm so}\left(|\Psi|, l=1\right) = \frac{\chi\Omega_L^2}{4\gamma^2}\left[-2\frac{|\Psi|^2}{S} + \frac{|\Psi|^4}{S^2}\right] . \qquad (6)$$

It corresponds to the conventional Ginzburg-Landau free energy in atomic BEC. The quadratic term modifies the potential U; the quartic term is now positive.

In the language of BEC, this means that, with increasing the density of Bose condensate, the originally attractive interaction between magnons should spontaneously become repulsive when the critical magnon density $n_c = S(8-\sqrt{8})/7$ is reached. If this happens, the magnon BEC becomes stable and in this way the state with spontaneous coherent precession could be formed. This self-stabilization effect is similar to the effect of Q-ball, where bosons create the potential well in which they condense. However, such a self-sustaining BEC with originally attractive boson interaction has not been achieved experimentally in bulk ^{3}He-A, most probably because of the large dissipation, due to which the threshold value of the condensate density has not been reached.

Finally the fixed orientation of the orbital vector $\hat{\mathbf{l}}$ has been achieved in ^{3}He-A confined in aerogel. Silicon strands of aerogel play the role of impurities with local anisotropy along the strands. According to the Larkin-Imry-Ma effect, the random anisotropy suppresses the orientational long-range order of the orbital vector $\hat{\mathbf{l}}$; however, when the aerogel sample is deformed the long-range order of $\hat{\mathbf{l}}$ is restored. Experiments with globally squeezed aerogel [37] demonstrated that a uni-axial deformation by about 1% is sufficient for global orientation of the vector $\hat{\mathbf{l}}$ along the anisotropy axis. When magnetic field is also oriented along the anisotropy axis one obtains the required geometry with $l = 1$, at which the magnon BEC in ^{3}He-A becomes stable. The first indication of coherent precession in ^{3}He-A has been reported in [42, 43, 44] and confirmed in [45] by observation of a long induction decay signal.

Contrary to the unconventional magnon BEC in the form of HPD in ^{3}He-B, the magnon BEC emerging in the superfluid ^{3}He-A is in one-to-one correspondence with the atomic BEC. For $\mu > U$, the condensate density determined from equation $dF/dn = \mu$ continuously grows from zero as $n \propto \mu - U$.

For $l = 1$ the Ginzburg-Landau free energy acquires the standard form:

$$F = \int d^3r \left(\frac{|\nabla\Psi|^2}{2m} + (\omega_L(\mathbf{r}) - \mu)|\Psi|^2 + \frac{1}{2}b|\Psi|^4\right), \qquad (7)$$

where we have modified the chemical potential by the constant frequency shift:

$$\mu = \omega + \frac{\Omega_L^2}{2\omega} , \qquad (8)$$

and the parameter b of repulsive magnon interaction is

$$b = \frac{\Omega_L^2}{2\omega S} \tag{9}$$

At $\mu > \omega_L$, magnon BEC must be formed with density

$$|\Psi|^2 = \frac{\mu - \omega_L}{b} \ . \tag{10}$$

This is distinct from ^{3}He-B, where condensation starts with finite condensate density. Eq. (10) corresponds to the following dependence of the frequency shift on tipping angle β of coherence precession:

$$\omega - \omega_L = -\frac{\Omega_L^2}{2\omega} \cos\beta \ . \tag{11}$$

The final proof of the coherence of precession in ^{3}He-A in aerogel was the observation of the free precession after a pulsed NMR and also after a switch off the CW NMR [45]. In conclusion, in contrast to the homogeneously precessing domain (HPD) in ^{3}He-B, the magnon Bose condensation in ^{3}He-A obeys the standard Gross-Pitaevskii equation.

Indeed, there is one important problem. The aerogel plays a role of impurities, which may be distributed very inhomogeneously. As a result the BEC state may be not a global state but a local state [46]. The example of HPD state in inhomogeneous aerogel is presented in Fig. 2.

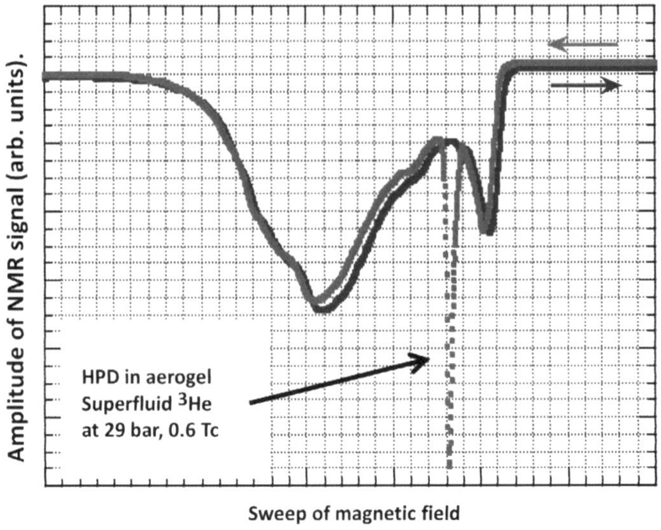

Figure 2. The HPD signal, observed in superfluid ^{3}He-B immersed in inhomogeneous aerogel. The HPD is formed only in a local place, which position was confirmed by an experiments at different gradient of magnetic field

The similar local states have been observed in ^{3}He-A and in HPD2 in aerogel. At pulsed NMR a few independent local BEC states are created, which induction signals are beating. The example of signal from a few BEC states of HPD2 are shown in Fig. 3.

Definitely, there are the gradient of magnetic ordering between the local BEC states, which leads to a relaxation due to spin diffusion current [44]. The relaxation may increase also due

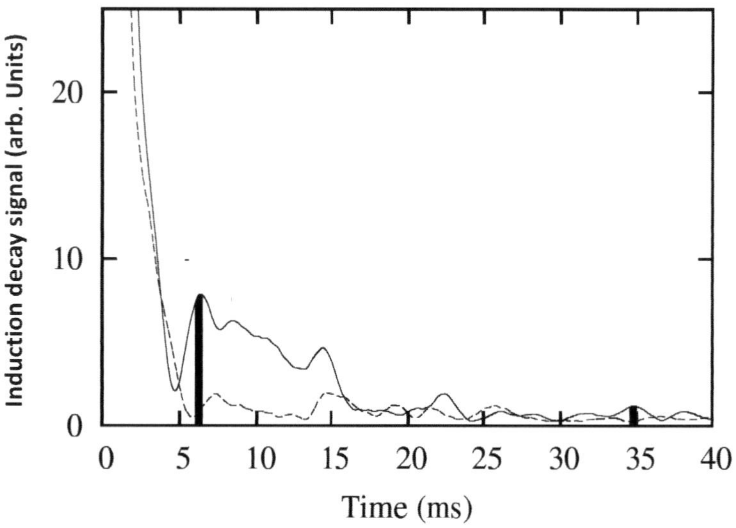

Figure 3. The beating of an induction signals from a few HPD2 states.

to local frequency shift owing to a local concentration of impurities. The topological defects may play an important role for magnetic relaxation as a sink region of magnons energy. All this demonstrate a great importance of the quality of the samples for observation of magnon BEC states. This problem becomes even more important for a magnon BEC in solid antiferromagnets, where the problem of impurities play a crucial role.

3. BEC in solid antiferromagnets

The magnon BEC in terms of coherent spin precession has been discovered in a monocrystals of $CsMnF_3$ and $MnCO_3$ [47]. The dynamic properties of coupled nuclear-electron precession in these antiferromagnets is very similar to a ^{3}He-A in aerogel [48]. It also shows the dynamical frequency shift (pulling). Magnetic systems with pulling were studied intensively in 70th of previous century [49]. This phenomenon takes place in the case when the frequencies of NMR and antiferromagnetic resonance are comparable. The typical system with pulling is the easy plain antiferromagnets with ions of Mn. The NMR frequency of Mn^{55} in hyperfine field is very high, about 600 MHz, and the antiferromagnetic resonance frequency in easy plain and cubic antiferromagnets can be very low. Due to coupling through the hyperfine field, the two modes of resonance appear, low frequency quasi NMR mode (NEMR) and high frequency quasi antiferromagnetic resonance mode (ENMR). The properties of magnetic subsystems change significantly due to the coupled precession. The hyperfine gap appears in a spectrum of antiferromagnetic spin waves. The quasinuclear spin waves appear, with the antiferromagnetic length of coherence. In other words, the nuclear magnetic system gets some properties of magnetically ordered system.

The pulse or CW magnetic resonance can move the spin system out of equilibrium. This deviation can be described in the language of elementary excitation, the spin waves (magnons). Magnons may condense very fast to a new equilibrium state. The density of non-equilibrium magnons depends on the angle of magnetization deflection β:

$$N = \frac{m - m_z}{\hbar} = \frac{\chi H}{\hbar\gamma}(1 - \cos\beta) \ . \tag{12}$$

The frequency dependence for NEMR mode in $CsMnF_3$ and $MnCO_3$ is very similar to that of superfluid ^{3}He-A (see Eq. 11):

$$\Omega = \omega_n - \omega_p \frac{m_z}{m_0} = \omega_n - \omega_p \cos\beta, \tag{13}$$

where ω_n and ω_p are the non-shifted NMR frequencies in hyperfine field and pulling, m_0 is the equilibrium nuclear magnetization and m_z is its projection on a hyperfine field. It means that both systems may have very similar properties. The condition of dynamic stability of coherent precession [10, 12] for NEMR mode in $CsMnF_3$ and $MnCO_3$ are fulfilled.

$$\frac{\partial\Omega}{\partial\cos\beta} < 0, \tag{14}$$

Owing this similarity, the magnon BEC observation in this systems was suggested [48]. The experiments demonstrated the formation of magnon BEC as by CW NMR technics [47] so by pulsed NMR experiments [50]. The last one was performed by a new techniques, developed in Ref. [45]. In this method the BEC is formed by a very long RF pulse. In this case the single BEC state is created. After the switching off of the RF pulse one can see a nice induction decay signal without beats. The details of experiments with $CsMnF_3$ and $MnCO_3$ one can find in Ref. [51, 52, 53]. Indeed, the length of this signal strongly depends on the quality of the crystal. The signal is 10 times longer in a new sample of better quality [54]. Definitely, the quality of the sample plays a crucial role for observation of a BEC in solid antiferromagnets.

It is important to point out that a similar experiment was described in Ref. [55]. The qualitative behavior of the signal was very similar. Indeed, it is difficult to figure out all parameters of the experiments, described there. To explain those results authors have used the scenario of saturation (heating) of the nuclear subsystem under a RF pumping. This scenario strongly contradicts the results of our experiments. The nuclear spin subsystem in these antiferromagnets is paramagnetic and after deflection should be dephased due to inhomogeneity of a superfine field. Furthermore, the dynamics of deflected magnetization should be described by Bloch equation. It means that the nuclear system should thermalize in a time T_2 and the transverse component should vanish. In our experiments there was found that it is not a case! In CW NMR experiments the deflected nuclear magnetization precess homogeneously at constant temperature and the NMR line does not saturate. The explanation can be found in the book [56]. This phenomenon was also considered in [57]. Two different scales for inhomogeneity were considered: the microinhomogeneous broadening for a distance smaller than the length of coherence of Suhl-Nakamura interaction (about 10^4 of the lattice size), and the macroscopic broadening for a longer distances. The microscopic broadening can be suppressed in the case when pulling is much bigger than a broadening. As a result the antiferromagnetic magnetization precess coherently at a short distance and force the nuclear subsystem precess coherently. For the longer distance it is not the case. It means that the nuclear subsystem shows the properties of ordered one, like the superfluid ^{3}He-A. Even the coherence lengths are comparable. The magnetization for this mode of resonance follows the Landau-Lifshits scenario. At the RF pumping it deflects but the module of magnetization does not change.

This work was done at the International Laboratory of the Magnetic Superfluidity and Non-linear Magnetic Resonance, Kazan Federal University.

References

[1] Borovik-Romanov A S, Bunkov Yu M, Dmitriev V V and Mukharskiy Yu M 1984 *JETP Lett.* **40**, 1033
[2] Borovik-Romanov A S, Bunkov Yu M, Dmitriev V V, Mukharskiy Yu M and Sergatskov D A 1989 *Phys.Rev.Lett.* **62**, 1631
[3] Borovik-Romanov A S, Bunkov Yu M, de Waard A, Dmitriev V V, Makroczyova V, Mukharskiy Yu M and Sergatskov D A 1988 *JETP Lett.* **47**, 478

[4] Borovik-Romanov A S, Bunkov Yu M, Dmitriev V V, Mukharskiy Yu M and Sergatskov D A 1990 *Physica B* **165**, 649
[5] Bunkov Yu M, Dmitriev V V and Mukharskiy Yu M 1986 *JETP Lett.* **43**, 168
[6] Bunkov Yu M, Dmitriev V V and Mukharskiy Yu M 1992 *Physica B* **178**, 196
[7] Kupka M,Skyba P, 2012 *Phys. Rev. B* **85**, 184529
[8] Bunkov Yu M 2004 *J. of Low Temp. Phys.* **135**, 337
[9] Bunkov Yu M 1995 *Spin Supercurrent and Novel Properties of NMR in* ^{3}*He*, (Prog. Low Temp. Phys. **14**, 69) ed W Halperin, (Amsterdam: Elsevier)
[10] Bunkov Yu M and Volovik G E 2008 *J. of Low Temp. Phys.* **150**, 135
[11] Bunkov Yu M and Volovik G E 2010 *J. Phys.: Condens. Matter* **22,** 164210
[12] Bunkov Yu M 2009 *J. Phys.: Condens. Matter* **21**, 164201
[13] Giamarchi T, Rüegg Ch and Tchernyshyov O 2008 *Nature Physics* **4**, 198
[14] Kagan Yu *et al.* 2007 *Phys. Lett. A* **361**, 401
[15] Bunkov Yu M and Volovik G E 2013 *Spin superfluidity and magnon BEC* (Novel Superfluids) ed. K. H. Bennemann and J. B. Ketterson (Oxford: University press)
[16] Bunkov Yu M 2005 *J. of Low Temp. Phys.* **138**, 753
[17] Bunkov Yu M and Volovik G E 2007 *Phys. Rev. Lett.* **98**, 265302
[18] Coleman S R 1985 *Nucl. Phys. B* **262**, 263
[19] Enqvist K and Mazumdar A 2003 *Phys. Rept.* **380**, 99
[20] Kusenko A, Kuzmin V, Shaposhnikov M and Tinyakov P G 1998, *Phys. Rev. Lett.* **80**, 3185
[21] Enqvist K and Laine M 2003 *JCAP* **0308** 003
[22] Bunkov Yu M, Dmitriev V V, Mukharskiy Yu M, Nyeki J and Sergatskov D A 1989 *Europhys. Lett.* **8**, 645
[23] Bunkov Yu M, Lvov V S and Volovik G E 2006 *JETP Lett.* **83**, 530
[24] Bunkov Yu M, Lvov V S and Volovik G E 2006 *JETP Lett.* **84**, 289
[25] Pitaevskii L and Stringari S 2003 *Bose-Einstein condensation* (Oxford: Clarendon Press)
[26] Bunkov Yu M, Fisher S N, Guenault A M and Pickett G R 1992 *Phys. Rev. Lett.* **69**, 3092
[27] Bunkov Yu M, Fisher S N, Guenault A M, Pickett G R and Zakazov S R 1994 *Physica B* **194**, 827
[28] Chen A S, Bunkov Yu M, Godfrin H, Schanen R and Scheffer F 1998 *J. Low Temp. Phys* **110**, 51
[29] Chen A S, Bunkov Yu M, Godfrin H, Schanen R and Scheffer F 1998 *J. Low Temp. Phys* **113**, 693
[30] Cousins D J, Fisher S N, Gregory A I, Pickett G R and Shaw N S 1999 *Phys. Rev. Lett.* **82**, 4484
[31] Autti S, Bunkov Yu M,Eltsov V B, Heikkinen P J, Hosio J J, Hunger P, Krusius M and Volovik G E 2012 *Phys. Rev. Lett.* **108**, 145303
[32] Bunkov Yu M, Collin E and Godfrin H 2005 *J. Phys. Chem. Solids* **66**, 1325
[33] Halperin W P,Choi H, Davis J P and Pollanen J 2008 *J. Phys. Soc. Jpn.* **77**, 111002
[34] Dmitriev V V, Zavjalov V V, Zmeev D E, Kosarev I V and Mulders N 2002 *JETP Lett.* **76**, 321
[35] Bunkov Yu M , Collin E, Godfrin H and Harakaly R 2003 *Physica B* **329-333**, 305
[36] Kunimatsu T, Matsubara A, Izumina K, Sato T, Kubota M, Takagi T, Bunkov Yu M and Mizusaki T 2008 *J. Low Temp. Phys.* **150**, 435
[37] Kunimatsu T, Sato T, Izumina K, Matsubara A, Sasaki Y, Kubota M, Ishikawa O, Mizusaki T and Bunkov Yu M 2007 *JETP Lett.* **86**, 216
[38] Elbs J, Bunkov Yu M, Collin E, Godfrin H and Volovik G E 2008 *Phys. Rev. Lett.* **100**, 215304
[39] Hunger P, Bunkov Yu M, Collin E and Godfrin H 2012 *J. Physics: Conference Series* **400**, 012019
[40] Fomin I A 1979 *JETP Lett.* **30**, 164
[41] Borovik-Romanov A S, Bunkov Yu M, Dmitriev V V and Mukharskiy Yu M 1984 *JETP Lett.* **39**, 469
[42] Bunkov Yu M and Volovik G E 2009 *JETP Lett.* **89**, 356
[43] Matsubara A, Sato T, Kunimatsu T, Izumina K, Kubota M , Mizusaki T and Bunkov Yu M 2009 *J. Phys.: Conf. Ser.* **150** 032052
[44] Sato T, Kunimatsu T, Izumina K, Matsubara A, Kubota M, Mizusaki T and Bunkov Yu M 2008 *Phys. Rev. Lett.* **101**, 055301
[45] Hunger P, Bunkov Yu M Collin E and Godfrin H 2010 *J. Low Temp. Phys.* **158**, 129
[46] Bunkov Yu M, Collin E, Godfrin H and Harakaly R 2003 *Physica B* **329**, 305
[47] Bunkov Yu M, Alakshin E M, Gazizulin R R, Klochkov A V, Kuzmin V V, Safin T R and Tagirov M S 2011 *JETP Letters* **94**, 68
[48] Bunkov Yu 2010 *Physics Uspekhi*, **53**, 843
[49] Borovik-Romanov A S, Bunkov Yu M, Dumesh B S 1977 *Physica B+C* **86** 1301
[50] Bunkov Yu M, Alakshin E M, Gazizulin R R, Klochkov A V, Kuzmin, V V, L'vov V S and Tagirov M S 2012 *Phys. Rev. Lett.* **108**, 177002
[51] Bunkov Yu M, Alakshin E M, Gazizulin R R, Klochkov A V, Kuzmin, Nizamutdinov A S,Safin T R and Tagirov M S 2011 *J. Phys.: Conf. Series* **324**, 012006

[52] Bunkov Yu M, Alakshin E M, Gazizulin R R, Klochkov A V, Kuzmin, Safin T R and Tagirov M S 2012 *J. Phys.: Conf. Series* **400**, 032001

[53] Alakshin E M, Bunkov Yu M, Gazizulin R R, Klochkov A V, Kuzmin, Rakhmatullin R M Sabitova A M , Safin T R and Tagirov M S 2013 Applied Magnetic Resonance **44**, 595

[54] To be published.

[55] Tulin V A 1969 *Sov. Phys. JETP* **28**, 431

[56] Kurkin M I and Turov E A 1990 *NMR in magnetically ordered materials and its applications* (Moscow, Nauka)

[57] Borovik-Romanov A S, Bunkov Yu M, Dumesh B S, Kurkin M I, Petrov M P andChekmarev V P 1984 *Sov. Phys. Uspekhi* **142**, 537

Moments distributions of single dye molecule spectra in a low-temperature polymer: Analysis of system ergodicity

T A Anikushina, A V Naumov

Institute for Spectroscopy, Russian Academy of Sciences, Troitsk, 142190, Russia
Moscow State Pedagogical University, Moscow, 119991, Russia

E-mail: anikushina@list.ru

Abstract. This article demonstrates the principal advantages of the technique for analysis of the long-term spectral evolution of single molecules (SM) in the study of the microscopic nature of the dynamic processes in low-temperature polymers. We performed the detailed analysis of the spectral trail of single tetra-tert-butylterrylene (TBT) molecule in an amorphous polyisobutylene matrix, measured over 5 hours at T = 7K. It has been shown that the slow temporal dynamics is in qualitative agreement with the standard model of two-level systems and stochastic sudden-jump model. At the same time the distributions of the first four moments (cumulants) of the spectra of the selected SM measured at different time points were found not consistent with the standard theory prediction. It was considered as evidence that in a given time interval the system is not ergodic

1. Introduction

One of the most significant areas of modern science and technology is the study of the properties of solid-state materials and structures composed of complex molecular systems (glasses, polymers and other). A significant amount of new information about the internal dynamics of solids can be obtained from the laser selective optical electronic-vibrational spectra of impurity centers (molecules, ions) embedded as spectral probes into a medium (matrix) transparent in the selected spectral range. The laser selective spectroscopy of impurity centers in solids is especially informative at cryogenic temperatures, when zero-phonon lines (ZPL) are reachable for observation (see [1] and references therein).

The possibilities of the methods of selective laser spectroscopy (persistent spectral hole burning and laser fluorescence line narrowing) are significantly limited because they provide data averaged over a large ensemble of impurity centers. The method of Single-molecule spectroscopy (SMS) completely eliminates this drawback. Single-molecule spectroscopy (SM), embedded as a spectral probe into a solid matrix, is a powerful tool for obtaining information about low-temperature dynamic processes in the local environment of these molecules [2]. In particular, SMS allows one to directly investigate the interaction of the electron transition impurity center with elementary excitations of a matrix of tunnel type - tunneling two-level systems (TLSs), (see [1-3] and references there).

In the present work we performed the analysis of the long-term temporal dynamics of the spectrum of a single chromophore molecules in the polymer matrix at cryogenic temperature, in order to clarify the applicability of the basic assumption of theories describing the spectral dynamics of such systems.

2. Evolution of spectra over time

As is well know from numerous studies, the low temperature (below several kelvins) dynamics of disordered solids is determined by the dynamics of specific low-energy elementary excitations - tunneling two-level systems (TLS)[4, 5]. Each (TLS) corresponds to the tunnel transitions of groups of atoms of a matrix between the two localized potential minima separated by a potential barrier. The interaction of the electronic transition of impurity chromophore molecules with active TLSs leads to spectral jumps of zero-phonon line (ZPL) of impurity chromophore molecule. As a result, the SM spectra changes during the time [3].

One of the most convenient and informative ways to observe the SM spectral evolution over a period of time is the method of multiple repeating detections of fluorescence excitation spectra within a selected spectral range [6]. The measurement of the SM spectra by recording their spectral trails has an important advantage; namely, it allows one to relatively, simply, and reliably identify the spectral lines and attribute them to individual molecules for further study [3, 7].

As a result of interaction of SM with an ensemble of TLSs spectra of different molecules have individual random shape and may consist of several peaks and stochastically alter with time (figure 1.b). In this case the description of the impurity system using only one parameter for the width of the SM spectral line is obviously not enough. To characterize and to explore the properties of impurity system, it was necessary to develop new approaches that adequately reflect the complex shapes of the spectra of single dye molecules.

To solve the above mentioned problem in the early 2000th it was suggested to describe the SM spectra within concepts of cumulants [8] and moments [9] of complex distributions. Moments and cumulants of various orders reflect the different peculiarities of complex (in general - multilines) SM spectral bands. In the case of a discrete distribution (in our case, the spectrum of single molecules), the first four central moments are expressed as follows:

$$M_1 = \frac{1}{\sum_i I_i} \sum_i I_i \omega_i ,\quad M_p = \frac{1}{\sum_i I_i} \sum_i I_i (\omega_i - M_1)^p \quad \text{p=2, 3, 4,} \tag{1}$$

where I_i is the signal amplitude of the data point of frequency ω_i.

In some cases, the description of complex distributions the concept of cumulants is more convenient, which are related to the central moments as follows:

$$\mathrm{k}_1 = \mathrm{M}_1 ,\quad k_2 = M_2 - M_1^2 . \tag{2}$$

As it has been shown in several studies, distribution of SM spectra cumulants, calculated on the basis of the standard TLS model and the stochastic model of sudden jumps [10], is described by the Levy statistics [9]. In this case the distribution of the first two cumulants of the spectra of single molecules P(k1), P(k2) [9] are of the form:

$$\mathrm{P}(\mathrm{k}_1) \sim \frac{\mathrm{z}_1}{\mathrm{k}_1^1 + \mathrm{z}_1^2} \text{ (Lorentz function), } P(k_2) \sim \frac{\mathrm{z}_{1/2}}{(\mathrm{k}_2)^{3/2}} \exp(-\frac{\mathrm{z}_{1/2}^2}{2\mathrm{k}_2}) \text{ (Smirnov function),} \tag{3}$$

where z_1 and $z_{1/2}$ coefficients are determined by the parameters of the impurity molecules and two-level systems. The very interesting property of this statistical description is that the ratio of parameters $z_1/z_{1/2}$ depends only on the general nature of dye-doped system (type of the dye-TLS interaction, isotropy and ergodicity of the system etc.).

In [11] it was shown that the distribution of the first and second cumulants of the spectra of 244 single molecules of tetra- tert -butylterrylene (TBT) in polyisobutylene (PIB) (420000) measured at T = 2 K, are satisfactorily described by the expressions (3), that is Levy statistics used to describe the measured spectra [8]. Moreover the value $z_1/z_{1/2}$ was obtained, which corresponds to the dipole-dipole type of interaction between SM and TLS, as well as a uniform spatial TLS distribution.

3. Approximation theory: the replacement of averages and the ergodic hypothesis

All the theories which describe the spectral characteristics of dye-doped disordered solid systems (e.g. for calculation of the photon-echo decay curves, shapes of spectral holes or spectra of single molecules), have quite complex mathematical apparatus which includes numerous averaging over various parameters. As result, in order to obtain analytical formulas different simplifications are applied based on some assumption about nature of system under study. For example, the in the theory of photon echo in low-temperature impurity glasses the correlation function $\Phi_{3\Phi Э}(\tau,t_w)$ is defined as follows [8]:

$$\Phi_{3\Phi Э}(\tau,t_w) = \left\langle \exp\left\{ -i\left[\int_0^{\tau} dt\omega(t) - \int_{t_w+\tau}^{t_w+2\tau} dt\omega(t) \right] \right\} \right\rangle_{stochastic,\varepsilon,A,J,config} \qquad (4)$$

Here, the angle brackets indicate averaging over all the two-level systems jumps histories (the so-called stochastic averaging, or averaging over the phase trajectories), over the TLS parameters and over the positions of the TLS (configuration averaging).

It should be noted that one of the essential conditions for the derivation of the analytical formulas was the validity of the ergodic hypothesis [12], namely, that the stochastic averaging over all phase trajectories can be replaced by averaging over time. Moreover, while analytically deriving the respective working formulas the integration sequence on various parameters was often changed [13]. In this context, the question remains how these approximations are valid. To date, the direct experimental data that shed light on this issue are lacking. The experiment provides only indirect information. Here we propose a method for studying this issue, which is to analyze the statistical regularities in the spectrum of the same single molecule which spectral dynamics were observed during a long time.

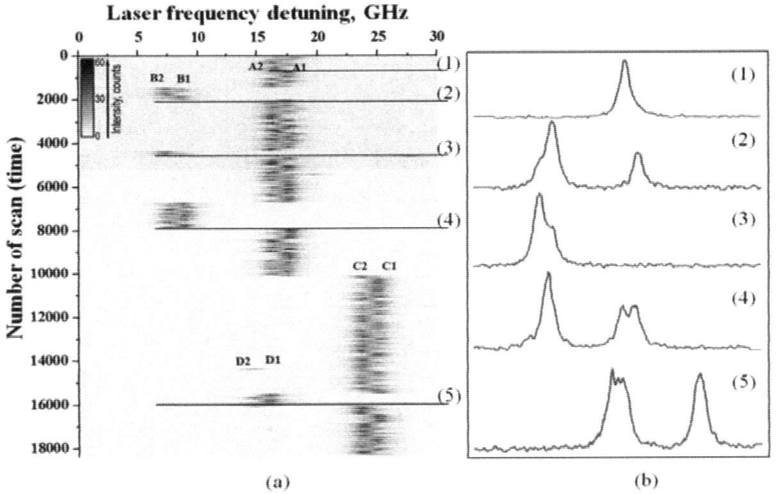

Figure 1. (a) Spectral trail of a single TBT molecule in PIB at T = 7 K. (b) Examples of spectra of the selected single TBT molecule as measured at different time points (shown by horizontal lines on (a)) with the selected integration time (120 seconds – 10 scans)

The spectral trail of the single molecule of tetra-tert-butylterrylene (TBT) in the thin (~0.5 μm) film of polyisobutylene (PIB) was measured over a long time (about 5 hours). Repeated measurement of the fluorescence excitation spectrum of the selected molecules was performed at T=7 K. Each scan was performed by a discrete tuning of the laser frequency in the 30 GHz range, number of frequency positions (steps) in scan was 500; exposition time per frequency point was 20 ms. Time for laser tuning back to the initial laser frequency was 2 seconds. By this way, 1536 scans were obtained (which corresponds to 18432 seconds). Two-dimensional plot of the measured spectral trail is shown in figure 1a.

As it is clearly resolved on the figure 1a. the ZPL of the selected TBT SM experiences correlated jumps between 8 spectral positions. It means that in the nearest environment of the analyzed chromophore there are three independent active TLS (we will call them as "strong" TLSs) which parameters are so specific that they cause ZPL jumps in frequency scale markedly out of ZPL width. First "strong" TLSs causes ZPL jumps like A1→A2, B1→B2, C1→C2; D1→D2. Transitions in the second "strong" TLS lead to the repeated ZPL jumps like A→B, C→D. Finally, third "strong" TLS excites the jump like A→C.

Of course, in the local environment of the observed impurity molecule there is a big number of active TLS, but most of them are either very slow, i.e. perform jumps extremely rarely and we cannot observe their dynamics in the spectrum of the SM during the selected period of time, or they are too far and lead to "small" (in frequency scale) jumps, which contribute to the overall broadening and shift of the spectral line.

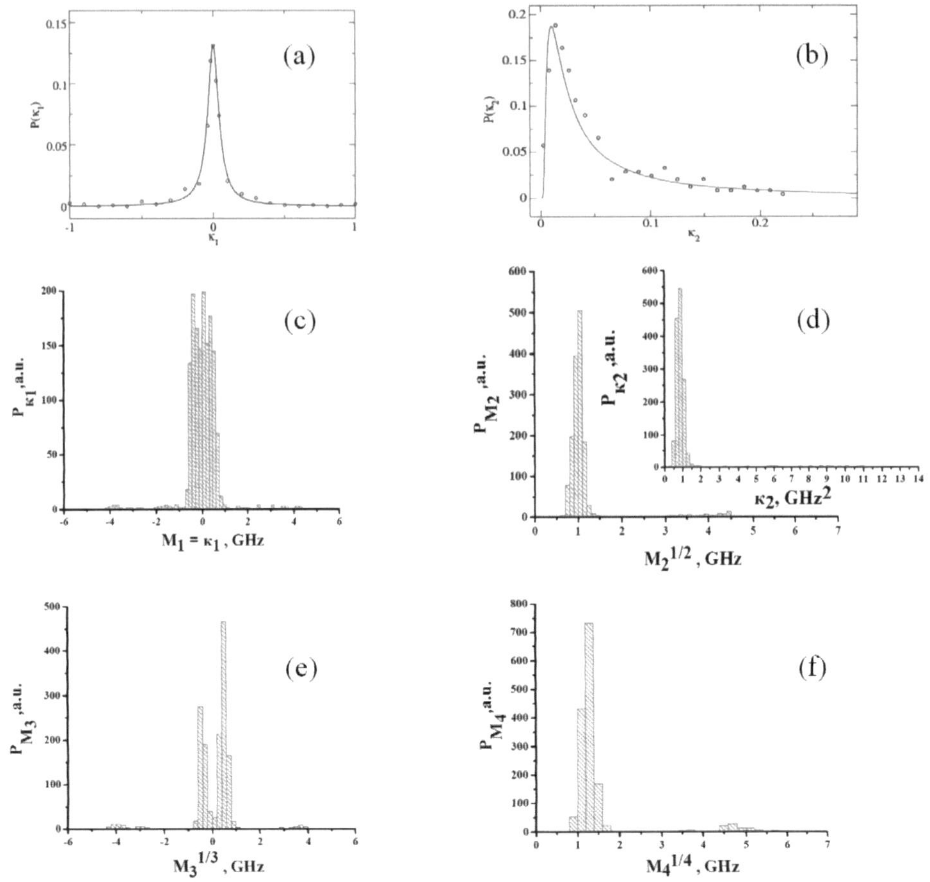

Figure 2. (a, b) Distribution of the 1st and 2nd cumulant spectra of single molecules, for an ensemble of 244 of TBT in PIB at T = 2K, approximated by a Lorentzian and Smirnov functions, respectively [11]. (c-f) Histograms of distributions of selected spectra of a single molecule of TBT in PIB at T = 7K, measured in different time points during the observation for about 5 hours.

In general, the data looks consistent with a model of the TLS. However, it is of interest to see how the system behaves over time and also make an attempt to analyze the applicability of the ergodic hypothesis in this case. Is our measurement time enough to verify its validity? For this we found the distribution of first four moments of the SM spectra measured in a different time points. Whole spectral trail was divided into sequential segments of 120 s. For each segment the integrated SM spectrum was found (see figure 1b). After that for each spectrum the first four moments (cumulants) of SM spectra were obtained using a special computer program. In addition width of a spectrum was defined by the approximation of a SM spectrum with Lorentz's function. In this way, a set of 1525 values or the width and the first four moments of SM spectra was found and the corresponding histograms were plotted (figure 2c-f).

Analysis of the distribution of points and comparison with the distributions obtained for the ensemble of single molecules in the same system TBT/PIB [11] leads to the following conclusions:

- The distributions of the first and second cumulants of spectra measured at different time points during the observation of the spectrum for 5 hours differ from the functions of the Lorentz and Smirnov (Eq. 3), respectively. This indicates that the Levy statistics is not applicable to describe the dynamics of the selected emitter in the realized time scale.
- There are some separate "tails" in the distributions of the second, third and fourth moments, which are associated with a strong splitting of the spectrum under the influence of a particular "strong" TLSs.

Based on above results, we can conclude that in a selected time interval (about 5 hours), the system is not ergodic. This fact must be taken into account in the theoretical analysis of the results in the field of spectroscopy of the impurity center, during the configuration and temporal averaging in the derivation of formulas describing the spectrum, a spectral hole shape, the photon-echo decay curves, etc.

Acknowledgments
This work was supported by the Russian Foundation for Basic Research, grant of the President of the Russian Federation (MD-465.2012.2, NSH-1049.2012.2). The experimental results were obtained and described earlier [14] in collaboration with Prof. Y.G.Vainer and Prof. L. Kador (University of Bayreuth, Germany).

References
[1] Naumov A V 2013 *Phys. Usp.* **56** 605
[2] Orrit M, Moerner W E 2011 *Physics and Chemistry at Low Temperatures* vol 1 (Singapore: Pan Stanford Publishing) pp 381-417
[3] Naumov A V, Vainer Y G 2009 *Phys. Usp.* **52** 298
[4] Anderson P W, Halperin, B I, Varma C M 1972 *Phil. Mag.* **25** 1
[5] Philips W A 1972 *J. Low. Temp. Phys.* **7** 351
[6] Ambrose W P, Moerner W E 1991 *Nature.* **349** 225
[7] Vainer Y G, Naumov A V, Bauer M, Kador L 2003 *Opt. Spectr.* **94** 864
[8] Naumov A V, Vainer Y G, Bauer M, Zilker S, Kador L 2001 *Phys Rev B.* **63** 212302
[9] Barkai E, Silbey R, Zumofen G 2000 *Phys. Rev. Lett.* **84** 5339
[10] Klauder J R, Anderson P W1962 *Phys. Rev.* **125** 912
[11] Barkai E, Naumov A V, Vainer Y G, Bauer M, Kador L 2003 *Phys. Rev. Lett.* **91** 075502
[12] Geva E, Skinner J L 1997 *J. Chem. Phys.* **107** 7630
[13] Sinai Ya G 1963 *Soviet Math. Dokl.* **4** 1818
[14] Vainer Y G, Naumov A V, Kador L 2008 *Phys Rev B.* **77** 224202

Journal of Physics: Conference Series **478** (2013) 012006 doi:10.1088/1742-6596/478/1/012006

Statistics for fluorescence photons of donor and acceptor molecules involved in non-radiative energy transfer (FRET)

A L Shchukina, I S Osad'ko, I Yu Eremchev

Institute for Spectroscopy of Russian Academy of Sciences, Troitsk 142190, Moscow, Russia

E-mail: a.schukina@inbox.ru

Abstract. Fluorescence of single molecular pairs involved in Förster Resonance Energy Transfer (FRET) is considered. Cases when either a donor of the electronic excitation energy or its acceptor has a non-fluorescent "dark" state are described. The influence of such dark states on energy transfer efficiency is analytically studied. Statistics for the number of donor/acceptor photons emitted per bin time is applied for these studies. Experimental distributions for photons of single emitters with "dark" states are presented, as well as their multivariate approximations with simulated distributions. It gives an example of how emitter parameters necessary for FRET efficiency calculations (transitions to and from the dark state) can be determined from the experimental statistics.

1. FRET: general

Single molecule spectroscopy, which emerged in 1990s [1-3], and superresolution techniques, which developed in 2000s [4-6], have naturally discovered new horizons in the research into the matter on the nanoscale level. Despite this, conventional optical techniques such as Förster Resonance Energy Transfer (FRET) are still in some cases the most suitable tool to probe processes undergoing there, especially in fragile biosystems [7].

The modern theory for FRET was presented by Th. Förster in 1948 [8]. In an act of FRET, a chromophore (donor), which has previously been electronically excited, transfers its excitation energy non-radiatively to another chromophore (acceptor) via dipole-dipole interaction. After this the donor is found in the ground state, and the acceptor – in the electronically excited one. In paper [8], the FRET rate was expressed via the spectral overlap between the donor emission and the acceptor absorption, the angle factor (a value depending on the angles between the donor/acceptor dipoles and the line connecting their centers) and the distance between two chromophores. The latter dependence allows using FRET as a "ruler" to measure distances on a nanometer scale.

Experimentally observable values are the donor/acceptor intensities. The quantity characterizing their relation is FRET efficiency E, which is by definition the ratio of the acceptor fluorescence intensity I_A to the summarized intensity of both emitters $I_A + I_D$:

$$E = \frac{I_A}{I_D + I_A}. \tag{1}$$

Journal of Physics: Conference Series **478** (2013) 012006 doi:10.1088/1742-6596/478/1/012006

2. Blinking fluorescence in FRET

According to (1), FRET efficiency E may vary from 0 (no energy transfer) to 1 (full energy transfer). However, expression (1) does not take account of possible intermittences in fluorescence of either of the emitters. Such intermittences, though, are often encountered due to various reasons in fluorescence of organic molecules [9], which are popular FRET donors and acceptors.

Figure 1 illustrates how fluorescence of a single quantum light emitter becomes blinking.

If a chromophore is excited with continuous-wave radiation, it jumps from the ground state *0* to the first excited state *1*. The contribution of higher excited states is negligible, so they are not considered. Level *2* in figure 1a is a long-lived non-fluorescent state. The rate of getting there is much less than the excitation and fluorescence rates, and so is the rate of leaving it. When a chromophore gets to state *2*, its fluorescence stops and resumes when it leaves that state. Thus, so-called off-states (or dark states) alternate with on-states (bright states) – its fluorescence is blinking (figure 1b).

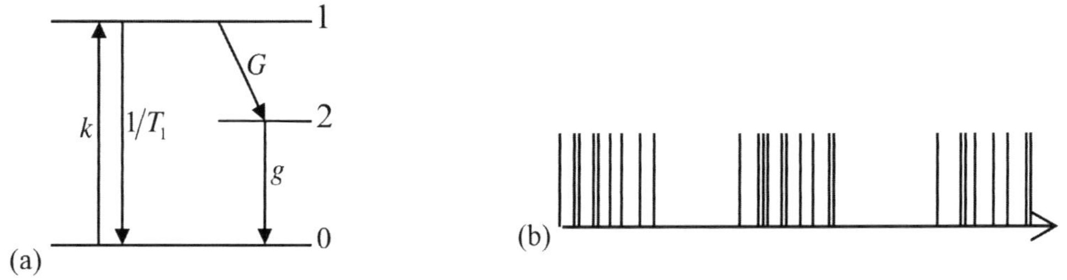

Figure 1. a) Three-level emitter. 0 – ground state, 1 – first excited state, 2 – "dark" state (off-state). k – excitation rate, $1/T_D$ – fluorescence rate, G and g – on-off and off-on transition rates. b) Instants of photon emission: blinking fluorescence of a three-level particle.

In FRET studies, blinking is usually neglected. Off-intervals are "cut" from the donor/acceptor fluorescence intensity records (trajectories), and only the remaining parts are used to calculate FRET efficiency. However, the present paper demonstrates a possibility to take into consideration full donor/acceptor trajectories with off-intervals.

3. Specific features of blinking in different types of donor-acceptor pairs

A donor and an acceptor can either be described by a 2-level electronic energy scheme (no off-state *2*, states *0* and *1* in figure 1b are only present, continuous fluorescence) or by a 3-level scheme (blinking fluorescence described above). Let us consider two types of donor-acceptor pairs: 3D-2A (3-level donor and 2-level acceptor) and 2D-3A (2-level donor and 3-level acceptor).

First of all, we have simulated donor and acceptor fluorescence for each type. The results are presented in figure 2.

Instants of donor and acceptor photon emission are shown with blue (upper) and red (lower) segments accordingly. The energy transfer rate F increases from the top to the bottom for each pair. The values of FT_D, where T_D is donor fluorescence characteristic time (fixed), are given on the left.

The present results were obtained with the help of the Monte-Carlo method based on rate equations for the populations of each energy level in a bimolecular system (donor + acceptor) (see Appendix A). The rates of all the transitions between various states of the system are considered constant, except for the energy transfer rate F, which is varied.

As can be seen, the presence of an off-state in one of the emitters affects the fluorescence of the other emitter as well. If a donor has an off-state, it may cause acceptor blinking, and vice versa [10].

In a 3D-2A pair, at medium energy transfer rates F (the third panel from the top of figure 2a) a 3-level donor blinks simultaneously with a 2-level acceptor, which has no dark state of its own. It happens because while a donor is in its dark state, it cannot transfer its energy to the acceptor, so the acceptor does not fluoresce either. When the transfer rate F becomes much bigger than the donor fluorescence rate $1/T_D$ (the bottom panel of figure 2a), both donor and acceptor fluoresce continuously, as the probability of the energy transfer overwhelms that of the donor on-off transition. Thus, simultaneous blinking is a sign of a donor off-state presence.

In a 2D-3A pair, on the contrary, blinking is not simultaneous, but alternating: either a donor or an acceptor has an off-state at a given moment (the fifth panel of figure 2b). When an acceptor gets into its dark state, the energy transfer stops competing with the donor fluorescence. That is why at large energy transfer rates F the donor fluorescence intensity increases from almost zero (at an acceptor in an on-state) to a considerable maximum (at an acceptor in an off-state). At lower F values (the third and the fourth panels of figure 2b), donor fluorescence is dual: its intensity alternates between two different values (the first is non-zero in this case).

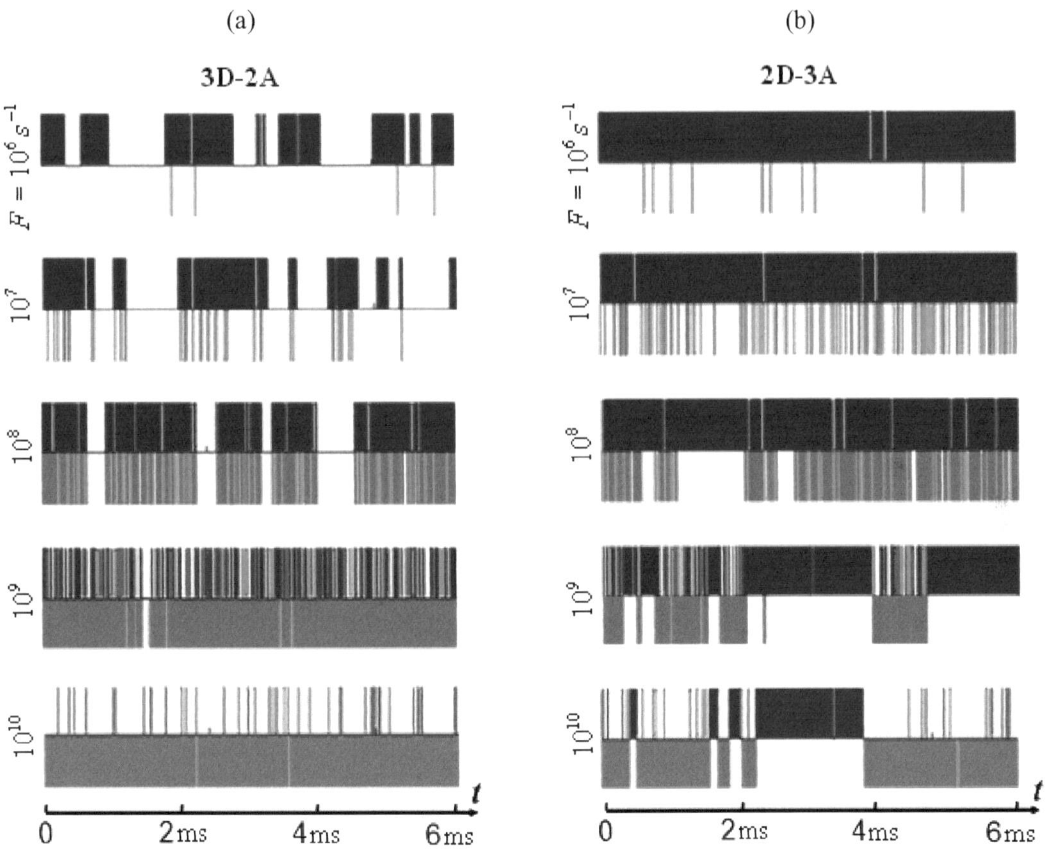

Figure 2. Instants of donor (upper blue segments) and acceptor (lower red segments) photon emission for pairs a) 3D-2A, b) 2D-3A. Parameters: $k = 10^6 s^{-1}$ (excitation rate), $1/T_D = 2 \cdot 10^8 s^{-1}, 1/T_A = 10^8 s^{-1}$ (donor/acceptor fluorescence rates), $G_D = 10^6 s^{-1}$, $g_D = 10^3 s^{-1}$, $G_A = 5 \cdot 10^5 s^{-1}$, $g_A = 500 s^{-1}$ (donor/acceptor on-off and off-on transitions).

This is how we gain the possibility to identify the case (3D-2A or 2D-3A pair) judging by the character of the donor/acceptor blinking [10].

4. Probability distribution functions for fluorescence photon numbers

An advantageous way to analyze the intensity trajectories like those presented in section 3 is to calculate the probability distribution functions for photon numbers emitted by a donor/acceptor at a given bin time. The method was developed in work [11]. According to this method, the time axis is divided into equal bin times (corresponding to those allowed by the time resolution in an experiment), and the number of photons emitted is counted at each bin time. Then the probability to observe a given photon number is calculated.

Figure 3 presents such photon number distributions for continuous (3a) and blinking (3b) fluorescence of a single emitter.

The distribution for continuous fluorescence (figure 3a) is close to a Poisson one. As it was shown in reference [12], they become indistinguishable when there is a big (about two orders of magnitude or more) difference between the excitation rate and the fluorescence rate, which is usually the case.

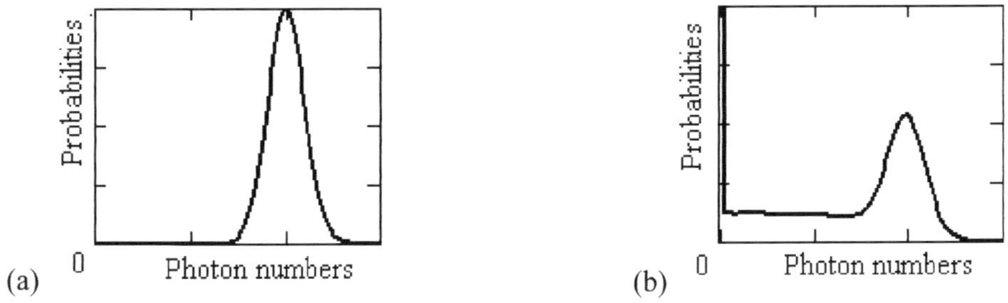

Figure 3. Photon number distributions (probabilities to register a given photon number at a bin time) for continuous (3a) and blinking (3b) fluorescence.

The distribution for blinking fluorescence (figure 3b) drastically diverges from the Poisson one. Figure 3b corresponds to a case when the bin time is comparable to the average on- and off-interval durations in blinking fluorescence. The presence of a plateau in the area of small photon numbers N, as well as a large probability to encounter zero photons (due to the possibility to cover a part of an off-interval with the observation interval) is a sign of blinking fluorescence.

Figure 4 shows some examples of photon number distributions for the intensity trajectories given in figure 2. Distributions for 3D-2A pairs were calculated both analytically and treating the computer simulated experiments (the two ways gave the same result). Distributions for 2D-3A pair were obtained with the help of computer simulation only. Dotted lines show Poisson distributions for comparison.

Having at our disposal distribution functions like those presented in figure 4, we can make a conclusion about the blinking or continuous character of donor/acceptor fluorescence – judging by the presence or absence of a plateau on the probability plot.

5. The contribution of background fluorescence

The model described above can be brought closer to real experiments by taking into account background fluorescence and bin time averaging. Figure 5 demonstrates simulated fluorescence of a 3-level donor shown previously in figure 2d (the second panel from the top) observed through bin times without background fluorescence (figure 5a) and with it (figure 5b).

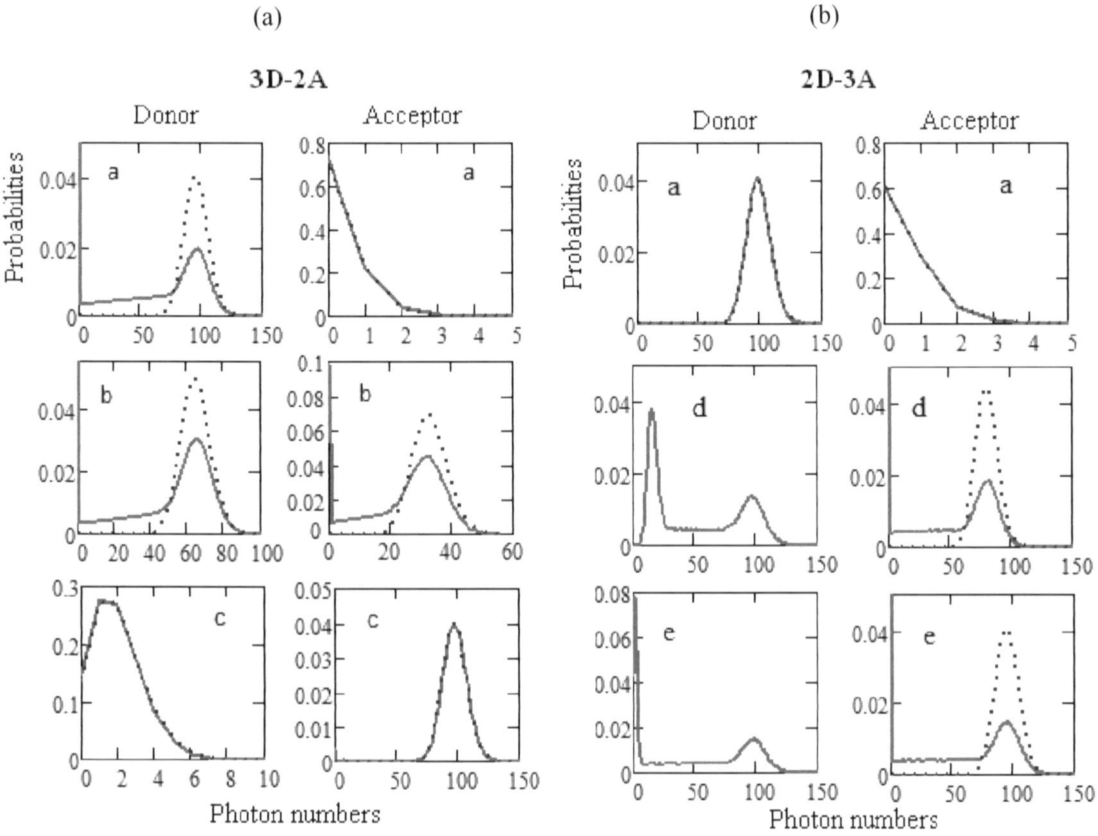

Figure 4. Photon number distributions for donor/acceptor fluorescence in 3D-2A and 2D-3A pairs. For parameters see figure 2. Bin time – 10ms. Letters a-e correspond to panels with different energy transfer rate F in figure 2.

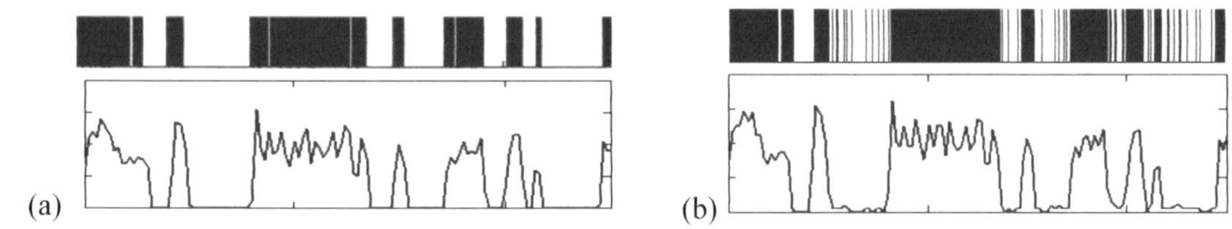

Figure 5. Simulated blinking fluorescence of a 3-level donor: separate photon emission instants (at the top) and photons counted within bin times (at the bottom): a) without background fluorescence; b) with background fluorescence of average intensity 20 times less than the on-interval donor intensity.

Though background fluorescence will blur the simultaneity/alternation of donor/acceptor blinking, it will still evidently allow making the necessary conclusions for choosing the right model – 3D-2A or 2D-3A. Off-intervals can be easily identified qualitatively from data like those in figure 5b. As for quantitative results, they can be obtained with the help of probability distribution functions for photon numbers (see section 4). Figure 6 presents such distributions for "pure" blinking fluorescence and that convolved with background fluorescence.

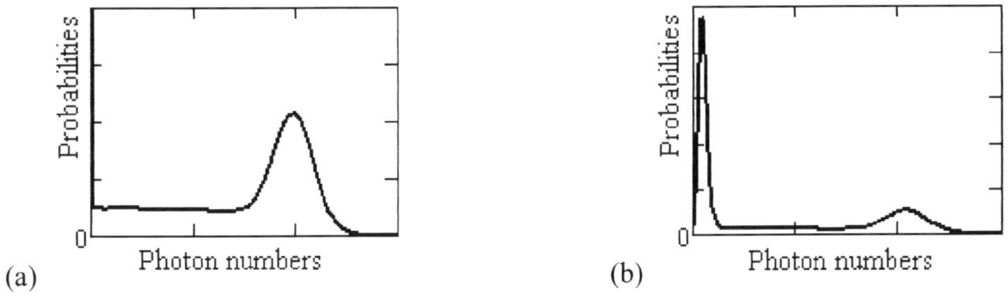

Figure 6. Photon number distributions a) for fluorescence of single 3-level emitter b) for fluorescence of 3-level emitter plus background fluorescence with Poisson statistics (intensity 20 times less than of the emitter during on-interval).

It was shown in reference [13] that when there are two sources of photons (say one is a fluorescing particle, and the other is background fluorescence), their joint signal should be treated with the help of the following formula:

$$w_{(2)}(N,T)=\sum_{i=0}^{N}\left(w_1(i,T)\cdot w_2(N-i,T)\right), \qquad (2)$$

that is, the probability $w_{(2)}(N, T)$ to register N photons from both sources at a bin time T is a convolution of the probabilities for each of the sources, w_1 and w_2. Thus, when background fluorescence is just "cut out", the distributions are distorted. So it is important to consider the convolution instead.

6. FRET efficiency for different types of donor-acceptor pairs

Formula (1) leads to a conventional expression of FRET efficiency E via donor fluorescence rate $1/T_D$ (T_D is the average lifetime of the first excited donor state in absence of an acceptor) and energy transfer rate F for a 2D-2A pair:

$$E=\frac{FT_D}{1+FT_D}. \qquad (3)$$

This expression can be generalized for each type of a donor-acceptor pair by replacing rate F in (3) with a function which contains the constants of all possible transitions in the system of the two emitters [10]. The corresponding formulas for 3D-2A and 2D-3A cases are given in Appendix B. They are derived with the help of the rate equations for the energy level populations (see Appendix A) in a stationary regime, i.e., setting the derivatives equal to zero, so that the donor/acceptor intensities become proportional to the populations of levels with an excited emitter.

The dependence of the FRET efficiency E on the energy transfer rate F for various types of donor-acceptor pairs is plotted in figure 6. 2D-2A (trivial case not) and 3D-2A pairs obey formula (2). Thus, the dark donor state does not influence the FRET efficiency. However, pairs with a 3-level acceptor do not exhibit full energy transfer even at large energy transfer rates (*E(F)* considerably lower than unity). So it can be concluded that an acceptor dark state hampers FRET efficiency [10].

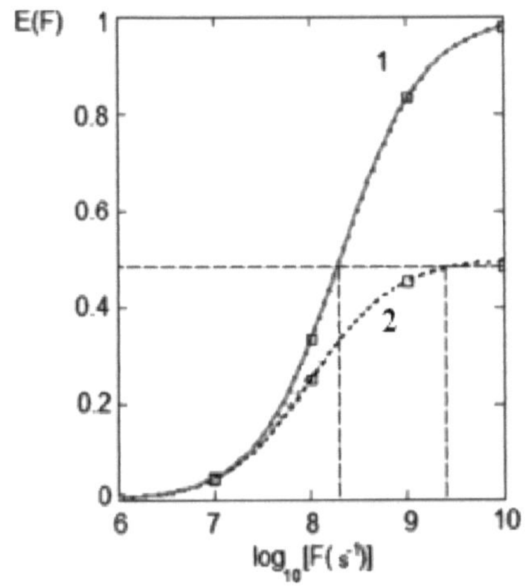

Figure 6. FRET efficiency *E(F)* depending on energy transfer rate *F* (logarithmic scale) for 3D-2A (1) and 2D-3A (2) pairs. Parameters for calculations are the same as in figure 2.

An experimentally established value of FRET efficiency, say, about 0.5 can thus correspond to two considerably different values of energy transfer rate (see dashed lines in figure 6). The difference can be of more than an order of magnitude, as is illustrated (the horizontal axis of figure 6 is in the logarithmic scale). The criterion, according to which the "real" value of the transfer rate should be chosen, is the simultaneity/alternation of donor/acceptor blinking. If they have simultaneous off-intervals, it is a 3D-2A pair, and we should follow line 1. If their off-intervals are alternating, it is a 2D-3A pair, and line 2 is the necessary one.

7. Experimental determination of transition rates in blinking fluorescence

To determine energy transfer rate *F* out of measurable FRET efficiency *E*, as section 6 and Appendix B show, we need to know the rates of excitation and on-off/off-on transitions of each molecule. The present section will show an example of how these parameters can be found from experimental photon number distributions of single organic molecules.

Single molecules fluorescence time tracks were obtained with a home-build luminescence microscope with epi-excitation scheme and multichannel registration system: cooled CCD with electronic multiplication (PCO Sensicam EM). Experiments were performed with Nikon high-numerical aperture microobjective. Exposition time was 30 ms. The sample (ATTO 610 dye molecules on a surface) was prepared from a water solution drop by spin-coating on a thin glassy plate (0.15 mm thickness). Single molecules were excited near their absorption peak by dye laser CR-599.

Single molecules images in each CCD frame (at a definite time point) were recognized and put in correspondence to different single molecules automatically by a custom computer program. It allows obtaining fluorescence time tracks for all single molecules in a microscope field of view. These tracks were used to calculate the experimental distributions of photon numbers per bin time (20 ms).

In order to obtain the necessary values of the parameters, the experimental distributions have been approximated with those simulated by the Monte-Carlo method. Firstly, "pure" distributions of a single blinking emitter like those presented in figure 3b were simulated. Then they were convolved with normal distributions (entailed by background fluorescence) in accordance with formula (2). Five parameters determine the shape of the convolved distribution in this case: for the blinking emitter – the excitation rate, the on-state lifetime and the off-state lifetime; for the background fluorescence – the average intensity (average number of photons per bin) and the width of its distribution. So, multivariate approximation of the experimental distribution was carried out to determine these five parameters. The method was a combination of Nelder-Mead algorithm and brute force (as there are many local minima of the difference between the experimental distribution and the simulated one). Figure 7 shows the results.

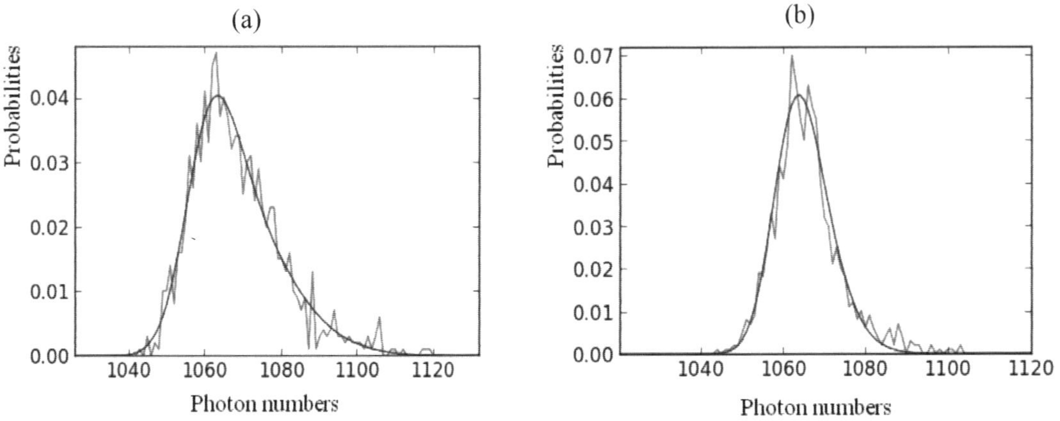

Figure 7. Two examples of experimental photon number distributions approximated by simulated ones. Parameters determined: a) $k = 1.3{\cdot}10^3\ s^{-1}$ (excitation rate), t_{on}=7.2ms (on-state lifetime), t_{off}=28.2ms (off-state lifetime), m=1060 (average background photons per bin), s=6.4 (distribution width for background fluorescence); b) $k = 7.8{\cdot}10^2\ s^{-1}$, t_{on}=4.0ms, t_{off}=14.3ms, m=1060, s=4.9. Bin time 30ms.

8. Conclusion

It has been shown that the type of one emitter involved in FRET (3-level or 2-level, i.e. producing blinking or continuous fluorescence) influences the fluorescence character of the other one. That is, a 3-level donor can make a 2-level acceptor blinking and vice versa. The difference is that in the first case (3D-2A) pair off-intervals occur simultaneously, while in the second one (2D-3A) they alternate.

Besides, it has been shown that a 3-level acceptor hampers FRET efficiency considerably and does not allow it to reach the value of 1 (full transfer) even at fast energy transfer. In this case, FRET efficiency depends not only on the energy transfer rate, but also on the rates of the transitions to and from the third “dark” acceptor state.

Finally, a way to determine these rates (included to the updated FRET efficiency dependence for a 2D-3A case) from experimental statistics has been described, and examples have been provided.

Acknowledgements

Authors acknowledge A.V. Naumov (Institute for Spectroscopy of Russian Academy of Sciences) for the custom program for fluorescence data retrieval. A.Shchukina acknowledges B.Elkin (Fraunhofer Institute for Interfacial Engineering and Biotechnology, Stuttgart, Germany) for stimulating discussions and ideas. Financial support from the Grants of the President of Russia (MK-6382.2012.2, MD-465.2012.2) is gratefully acknowledged.

Appendix A

Rate equations for two types of donor-acceptor pairs.

3D-2A:

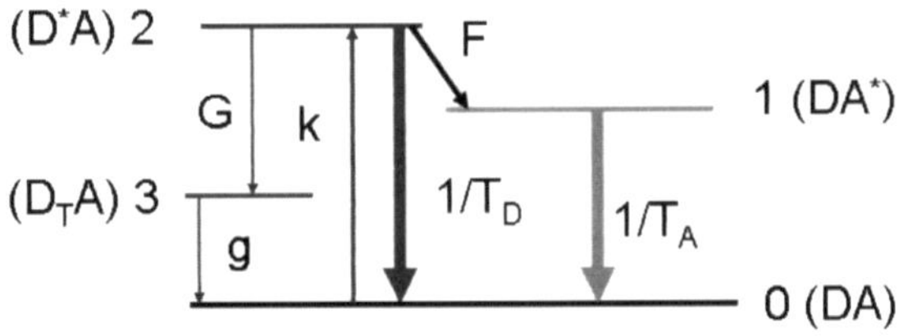

$$\dot{\rho}_0 = -k\rho_0 + \rho_1/T_A + \rho_2/T_D + g\rho_3,$$
$$\dot{\rho}_1 = -\rho_1/T_A + F\rho_2,$$
$$\dot{\rho}_2 = k\rho_0 - (F + G + 1/T_D)\rho_2,$$
$$\dot{\rho}_3 = G\rho_2 - g\rho_3.$$

D*, A* - excited donor/acceptor. Here and in what follows we neglect the level D*A* (both donor and acceptor are excited), as its population is diminishable. D_T – donor in a dark state. k – excitation rate, $1/T_D$ – fluorescence rate, G and g – on-off and off-on transition rates,

2D-3A:

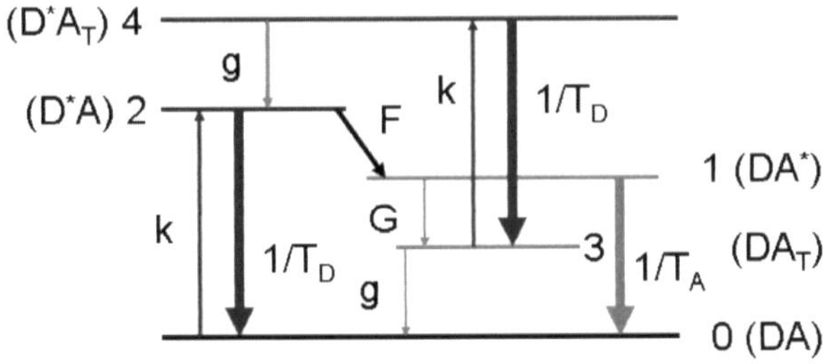

$$\dot{\rho}_0 = -k\rho_0 + \rho_1/T_A + \rho_2/T_D + g\rho_3,$$
$$\dot{\rho}_1 = -(G + 1/T_A)\rho_1 + F\rho_2,$$
$$\dot{\rho}_2 = k\rho_0 - (F + 1/T_D)\rho_2 + g\rho_4,$$
$$\dot{\rho}_3 = G\rho_1 - (k + g)\rho_3 + \rho_4/T_D,$$
$$\dot{\rho}_4 = k\rho_3 - (g + 1/T_D)\rho_4.$$

A_T – acceptor in a dark state

Appendix B

Formulas for FRET efficiency for various types of donor/acceptor pairs:

$$E(F) = \frac{\bar{F}(F)T_D}{1 + \bar{F}(F)T_D}$$

for 2D-2A and 3D-2A:

$$\bar{F}(F) = F$$

for 2D-3A:

$$\bar{F}(F) = \frac{F}{1 + GT_A + \frac{G}{g}\frac{FT_D kT_A}{(k+g)T_D + 1}}$$

References

[1] Moerner W E, Kador L 1989 Phys. *Rev. Lett.* **62** 21 2535
[2] Orrit M, Bernard J 1990 *Phys. Rev. Lett.* **65** 2716
[3] Ambrose W P, Basche Th, Moerner W E 1991 *J. Chem. Phys.* **95** 7150
[4] Dyba M, Jakobs S, Hell S W 2003 *Nat. Biotechnol.* **21** 1303
[5] Gustafsson M G L 2005 *Proc. Natl. Acad. Sci. USA* **102** 1308
[6] Betzig E, Patterson G H, Sougrat R, Lindwasser O E, Olenych S, Bonifacino J S, Davidson M W, Lippincott-Schwartz J, Hess H F 2006 *Science* **313** 1642
[7] Grecco H E, Verveer P J 2011 *Chemphyschem.* **12** 3 484
[8] Förster Th 1948 *Annalen der Physik* **6**(2) 55
[9] Orlov S V, Naumov A V, Vainer Yu G, Kador L 2012 *J. Chem. Phys.* **137** 194903
[10] Osad'ko I S, Shchukina A L 2012 *Phys. Rev. E* **85** 061907
[11] Osad'ko I S, Fedyanin V V 2009 *J. Chem. Phys.* **130** 064904
[12] Osad'ko I S 2005 *ZhETF* **128**(1) 77
[13] Osad'ko I S, Shchukina A L 2010 *J. Phys. Chem. C* **114**(23) 10349

Luminescence properties of CdSe nanocrystallites in cadmium alkanoate glasses

K A Magarian[1], V V Fedyanin[1,2], K R Karimullin[3,4], I A Vasilieva[1] and G V Klimusheva[5]

[1] Moscow State Pedagogical University, 29 Malaya Pirogovskaya Str., Moscow, 119992, Russia
[2] P. N. Lebedev Physics Institute of RAS, 53 Leninskii prosp., Moscow, 119991, Russia
[3] Institute for Spectroscopy of RAS, 5 Fizicheskaya Str., Troitsk, Moscow, 142190, Russia
[4] E.K. Zavoyski Kazan Physical-Technical Institute of RAS, 10/7 Sibirski trakt Str., Kazan, 420029, Russia
[5] Institute of Physics, NAS of Ukraine, 46 Prospect Nauki, Kiev, 03028, Ukraine

E-mail: xmagaros@gmail.com

Abstract. Quantum confinement effect and its influence on absorption and photoluminescence spectra of CdSe nanoparticles chemically synthesized in thermotropic ionic liquid crystal of cadmium octanoate are studied. It was found out that the spectral resolution grows with decrease of temperature, so it becomes possible to distinguish the band-edge luminescence from the bands associated with the presence of surface defects. Obtained spectra were analyzed theoretically and the energy diagrams for samples with 1.8 nm and 2.6 nm nanoparticles were calculated.

1. Introduction

Semiconductor nanocrystals (quantum dots, QD) of metal chalcogenides including cadmium selenide attract much attention for half a century, because these media possess vast potential capacity to solve various problems in different fields of activity including material sciences, electronics, biology and medicine. Properties of such compounds, especially their optical properties depend on the size of quantum dots, i.e. on dimensions of the spatial confinement area for electrons, holes and excitons movement [1-3]. Thus the opto-electronic properties of nanocrystals are defined by the quantum confinement effect [4, 5]. Local environment of nanocrystals and methods of their synthesis also influence these properties. Recently more and more complex matrices with anisotropic structure, oriental order and periodicity, such as mesomorphic polymers are used (e.g., [6]) to obtain new materials with appropriate characteristics. One of such new media are liquid crystal matrices of cadmium alkanoates composites with CdSe nanoparticles (NPs), whose absorption spectra and photoluminescence properties were investigated at room temperature in [7]. In the present paper we studied spectral and luminescence characteristics of CdSe quantum dots in cadmium alkanoate glasses in a broad temperature range.

2. Objects of investigation

We have investigated CdSe NPs, chemically synthesized in a thermotropic ionic liquid crystal phase of a cadmium octanoate at high temperature. In the cadmium alkanoate matrices [8], the semiconductor NPs have small size dispersion, they are stable and ordered in a layered smectic A matrix [9].

All cadmium octanoate composites with CdSe NPs were synthesized by T.A. Mirnaya`s group (V.I. Vernadsky Institute of General and Inorganic Chemistry, NAS of Ukraine). The samples of the glassy nanocomposites were prepared by the following method. Polycrystalline powder of synthesized composite was placed between flat substrates and then it was heated to the temperature of liquid crystalline mesophase. After that the sample was cooled rapidly, leading to creation of an anisotropic glass at room temperature [9]. Three types of samples were prepared, containing 1.0 nm, 1.8 nm and 2.6 nm CdSe NPs. Thickness of composite samples was about 10 - 30 μm. The nanoparticles concentration was 4 mol% in all samples.

3. Experiment

Absorption spectra for all samples at room temperature were measured using spectrophotometer Specord M400. Registration of fluorescence spectra and fluorescence excitation spectra was performed by means of the optical system based on the two monochromators MDR-72 with a linear dispersion of 4 nm/mm. The source of excitation light was a 1000W Xe-lamp. Fluorescence spectra were detected on different excitation wavelengths at room temperature, 77 K and 4.2 K.

It should be noticed that quantum yield was increased with the temperature decreasing to 77 K. With further decrease of a temperature to 4.2 K the fluorescence intensity continued to grow, though the spectral peaks positions stood the same compared with ones at 77 K.

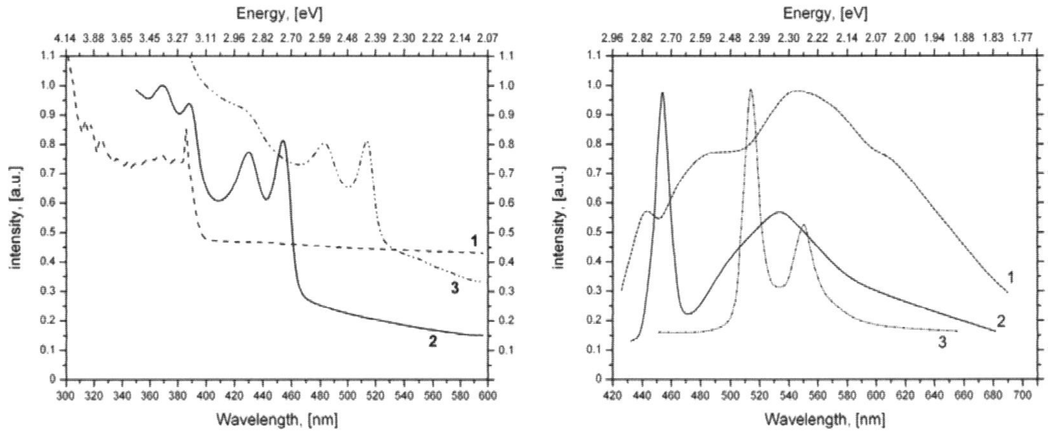

Figure 1. The absorption spectra (left-hand graph) and fluorescence spectra (right-hand graph) of nanocomposites at room temperature. NPs sizes and corresponding excitation wavelengths are: **1** – 1.0 nm, λ_{exc}=393 nm; **2** – 1.8 nm, λ_{exc}=387 nm; **3** – 2.6 nm, λ_{exc}=387 nm.

4. Results and discussion

In the absorption spectra there are several peaks that can be explained by the presence of NPs with flat form. The double-humped absorption spectra of CdSe NPs are more typical for platelet nanoparticles. In this case the absorption bands at 430 nm and 455 nm can be associated with electron transitions from light hole (LH) and heavy hole (HH) energy levels of valence band to the lowest energy level of conduction band [10]. The absorption spectra clearly demonstrate a size dependent quantum confinement effect. The absorption bands of the CdSe NPs are shifted to the high-energy region relatively to the band gap energy E_g = 1.7 eV of the CdSe bulk crystal [11]. Pure cadmium alkanoate matrices without quantum dots do not absorb light in a visible spectrum.

Both absorption and fluorescence spectra at room temperature consist of separate resonant exciton peaks for a samples with 1.8 nm and 2.6 nm NPs. Decreasing of the NPs size leads to a blue shift in a conjugate spectra. Exciton peaks are the most intensive in the fluorescence spectra (figure 1).

Decreasing of temperature to 77 K causes the following changes in spectra: at 77 K the fluorescence spectrum of the 1.0 nm NPs (figure 2) becomes more structured compared with the room temperature spectrum. There is a short-wave exciton peak with a half-width ~ 15 nm. Between the exciton peak in the fluorescence spectra and the most long-wave band in the fluorescence excitation spectra (that can be related to an exciton transfer) there is a Stokes shift ~ 50 nm. Relative intensity of the long-wave bands at 77 K is much greater compared with the room temperature measurements.

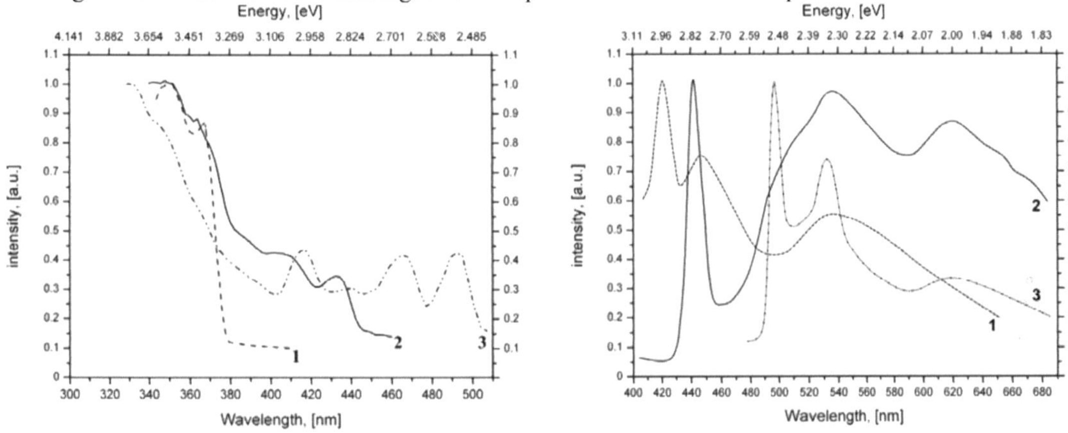

Figure 2. Fluorescence excitation spectra (left-hand graph) and fluorescence spectra (right-hand graph) of nanocomposites at 77 K. Sizes of NPs and corresponding excitation wavelengths are (left): **1** – 1 nm, λ_{reg}=538 nm, λ_{exc}=393 nm; **2** – 1.8 nm, λ_{reg}=535 nm, λ_{exc}=387 nm; **3** – 2.6 nm, λ_{reg}=532 nm, λ_{exc}=387 nm. Fluorescence excitation spectra were recalculated considering the distribution of the radiation source intensity.

The fluorescence excitation spectra show a blue shift between nanocrystal absorption bands and the band–gap of a bulk CdSe crystal which is about 1.82 eV at 80 K [12]. Further decrease of a temperature down to 4.2 K doesn't lead to any changes either in peaks position or in half-width values in fluorescence spectra of 1.8 nm and 2.6 nm NPs samples (figure 3).

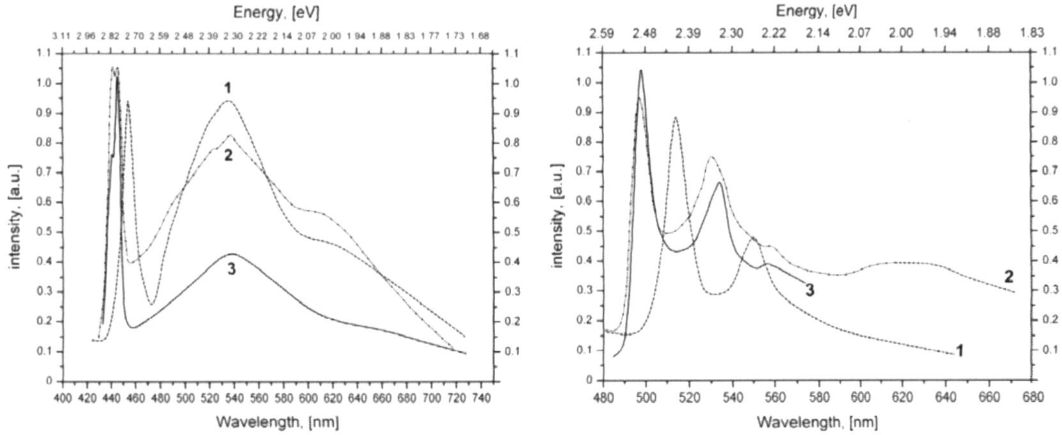

Figure 3. The fluorescence spectra of 1.8 nm CdSe QD (the same λ_{exc}=370 nm) (left-hand graph) and 2.6 nm. CdSe QD (the same λ_{exc}=430 nm) (right-hand graph): **1** – at room temperature; **2** – at 77 K; **3** – at 4.2K.

The energy diagrams were built based on the analysis of the obtained spectra for 1.8 nm and 2.6 nm NPs samples in cadmium alkanoate glasses (figure 4). Notes "0" and "1" in the figure indicate energy levels of size-quantization - hole energy level and electron level, respectively; "2" and "3" denote the levels for trapping carriers. L – is absorption energy.

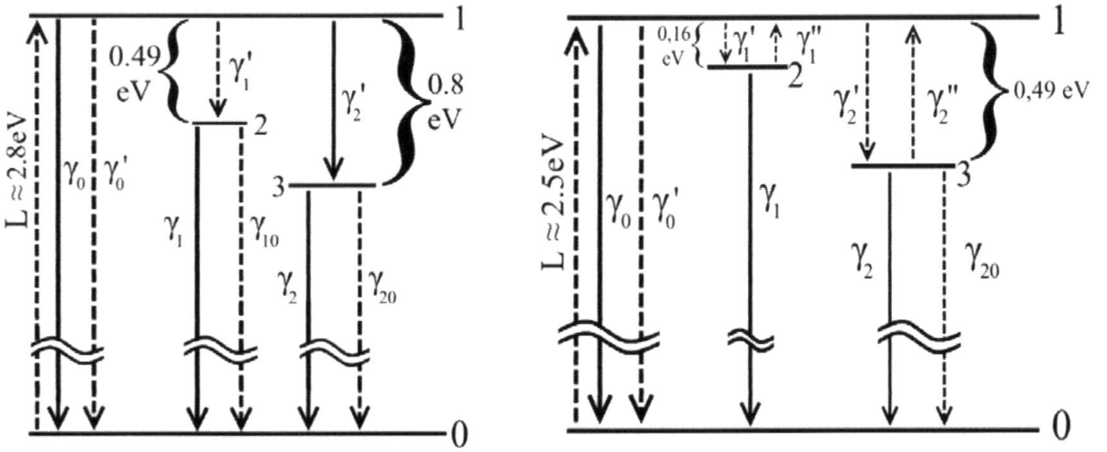

Figure 4. Schematic energy diagrams of 1.8 nm CdSe NPs (left-hand graph) and 2.6 nm CdSe NPs (right-hand graph) in cadmium alkanoate glasses. See the text for notations.

Transition 1-0 represents an exciton band in a fluorescence spectrum. Transitions 2-0 and 3-0 are corresponded to the long-wave bands. Rates γ_0 and γ_0' characterize radiative and non-radiative recombination of electrons and holes in the size-quantization levels, respectively. Besides the band-edge transition there are two processes of trapping non-equilibrium carriers from the size-quantization levels to the shallow traps: 1-2 and 1-3 with corresponding rates γ'_1 and γ'_2 for electrons.

Assuming n, n_1, and n_2 as the average number of the electron-hole pairs in the appropriate states per one NP, the balance equations for 1.8 nm CdSe NPs will be following:

$$\begin{aligned} \dot{n} &= L - (\gamma_0 + \gamma_0' + \gamma_1' + \gamma_2')n \,, \\ \dot{n}_1 &= \gamma_1' n - (\gamma_1 + \gamma_{10})n_1 \,, \\ \dot{n}_2 &= \gamma_2' n - (\gamma_2 + \gamma_{20})n_2 \,. \end{aligned} \qquad (1)$$

Solving equations (1) for the case when $\dot{n}_i = 0$, the spectral bands intensities can be found as: $I=\gamma_0 n$, $I_1=\gamma_1 n_1$, $I_2=\gamma_2 n_2$. One can obtain from the equations (1) that the excitation intensity L doesn't influence the relative fluorescence intensities (I / $I+I_1+I_2$ etc). Other parameters are expressed in the units of γ_0 of the main transition. Radiative rates $\gamma_0 = \gamma_1 = \gamma_2 = 1$. Non-radiative rates were used in the following form: Const$\cdot e^{-0,05/kT}$ in accordance with the work [13], where 5 nm CdSe QDs were studied. Values of the rates had been selected so as the measured spectrum was in close agreement with one calculated using the equations (1). Probability of γ_2'' transition is many orders smaller than the others, so it wasn't considered in calculations. All spectra were approximated by Gaussian functions.

Table 1. Values of the non-radiative constants for fluorescence spectra of 1.8 nm and 2.6 nm NPs.

QD size (nm)	λ_{exc} (nm)	γ_0' ($\times e^{0.05/kT}$)	γ_1'	γ_2'	γ_1'' ($\times e^{0.05/kT}$)	γ_{10} ($\times e^{0.05/kT}$)	γ_{20} ($\times e^{0.05/kT}$)
1.8	387	170	4.14	11.52	-	2	50
2.6	410	26	1.50	2.70	400	-	50

The optimized values of non-radiative constants for fluorescence spectra of the samples containing 1.8 nm and 2.6 nm CdSe NPs are given in the table 1.

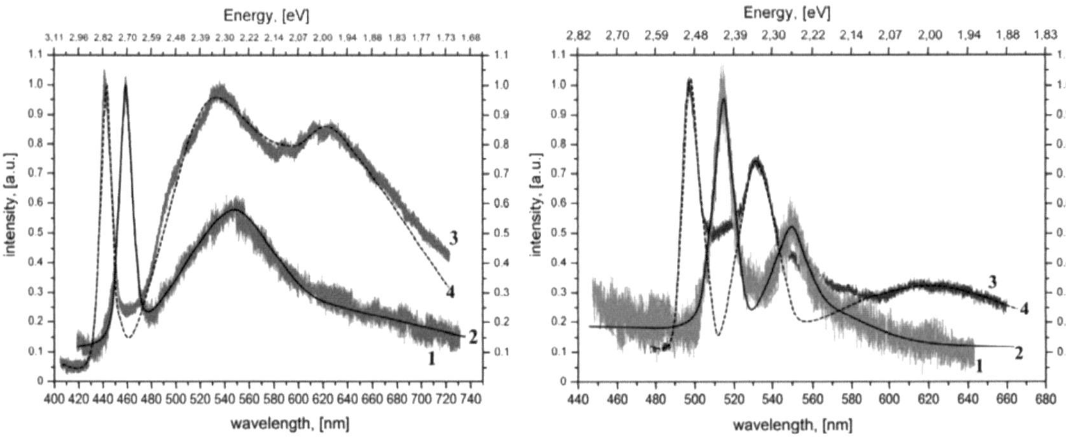

Figure 5. Luminescence spectra of 1.8 nm (left) and 2.6 nm (right) QD nanocomposites, measured: **1** – at room temperature and **3** – at 77 K, and calculated using the equations (1): **2** – at room temperature, **4** – at 77 K.

The examples of calculated spectra of 1.8 nm and 2.6 nm CdSe nanocrystals at room temperature and 77 K are shown in the figure 5.

5. Conclusion

We have measured absorption and luminescence spectra of CdSe NPs with average size of 1.0 nm, 1.8 nm and 2.6 nm which are chemically synthesized in thermotropic ionic cadmium octanoate liquid crystal. Anisotropic glassy composites with CdSe NPs are obtained by rapid cooling from mesophase to the room temperature. Obtained spectra were analyzed theoretically and the model of electron levels of CdSe nanoparticles is offered. Developed theoretical approach satisfies experimental data with quite high accuracy to a first approximation. In the future, we are going to perform a more detailed analysis of these spectra and consider their relation with the structure of the CdSe NPs.

Acknowledgments

The research is partially supported by the Ministry of Education and Science of the Russian Federation, contract # 14.B25.31.0007 and by the Russian Foundation for Basic Research, project # 12-02-31381.

References

[1] Norris D J, Efros Al L, Rosen M and Bawendi M G 1996 Size dependence of exciton fine structure in CdSe quantum dots *Phys. Rev.* B **53** 16347-54

[2] Michalet X, Pinaud F, Lacoste T D, Dahan M, Bruchez M P, Alivisatos A P and Weiss S 2001 *Single Mol.* **2** 261

[3] Parak W J, Gerion D, Pellegrino T, Zanchet D, Micheel C, Williams S C, Boudreau R, Le Gros M A, Larabell C A and Alivisatos A P 2003 *Nanotechnology* **14** R15

[4] Klimov V 2004 *Semiconductor and metal nanocrystals* New York, Marcel Dekker Inc.

[5] William Yu W, Lianhua Qu, Wenzhuo Guo, Xiaogang Peng 2003 Experimental Determination of the Extinction Coefficient of CdTe, CdSe, and CdS nanocrystals *Chem. Mater.* **15(14)** 2854-60

[6] Shatalova A M, Kutergina I Yu, Derikov Ya I, Shandryuk G A and Talroze R V 2012 Matrices based on acrylic liquid-crystalline copolymers for the design of composites with quantum dots *Polymer Science Series* B **54** 533-41

[7] Klimusheva G, Dmitruk I, Mirnaya T, Tolochko A, Bugaychuk S, Naumenko A, Melnik D and Asaula V 2013 Monodispersity and ordering of semiconductor quantum dots synthesized in ionic liquid crystalline phase of cadmium alkanoates *Liquid Crystals* **40** 980-8
[8] Mirnaya T A and Volkov S V in: Rogers R D, Seddon K R and Volkov S V (Eds.) 2002 Green industrial applications of ionic liquids, *Kluwer Academic Publishers* London 439-56
[9] Lyashchova A, Fedorenko D, Garbovskiy Yu, Klimusheva G, Mirnaya T and Asaula V 2013 Strong Thermal optical nonlinearity caused by CdSe nanoparticles synthesized in smectic ionic liquid crystal *Liquid Crystals* **40** 1377-82
[10] Ithurria S, Tessier M D, Mahler B, Lobo R P S M, Dubertret B and Efros Al L 2011 Colloidal nanoplatelets with two-dimentsional electronic structure *Nature materials* **10** 936 – 41
[11] Gusev A I 2005 *Nanomaterials, nanostructures, nanotechnology* Moscow: Fizmatlit
[12] Karpov S V and Mikushev S V 2010 Electron-hole excitations in CdSe quantum dots under strong and intermediate confinement conditions *Physics of the Solid State* **52** 1750-56
[13] Katsaba A V *et al* 2013 Characterization of defects in colloidal CdSe nanocrystals with the use of modified thermally stimulated luminescence method. *Physics and Technique Semiconductors* **47** 1339-43

Control of the index of refraction in optically dense medium

K A Barantsev and A N Litvinov

St. Petersburg State Polytechnical University, Saint Petersburg 195251, Russia

E-mail: kostmann@yandex.ru

Abstract. In this work we investigate conditions of emergence and destruction of the electromagnetically induced transparency effect in the optically dense medium of atoms which are excited along closed contour of excitation by the three frequency radiation. We found the conditions under which index of refraction of the medium acquires spatial quasi-periodic dependence. It is possible to control this dependence by variation of the initial phases and intensities of the electromagnetic fields.

1. Introduction

The mediums with spatial periodical dependence of index of refraction, which is capable of the Bragg diffraction of light have the properties of photonic crystals [1-2]. However, variation of the optical properties of such crystals, specifically variation of the index of refraction in different layers, is possible only at growing of a crystal. At the same time increased interest is attracted to the mediums in which it is possible to induce the quasi-periodic structure of the resonance index of refraction by electromagnetic fields under the electromagnetically induced transparency (EIT) effect [3].

This work is devoted to investigation of emergence of stationary quasi-periodic structure of the index of refraction in the medium of cold atoms which interact with three electromagnetic fields, propagating collinear. In this calculation we suppose that the microwave field inside the medium is known only. It interacts with atoms weakly because the magnetic dipole moment of microwave transition is much smaller than electric dipole moments of the optical transitions. We suppose that two electromagnetic fields of optical range are unknown in the thickness of the medium and calculate them by the self-consistent solution of the Maxwell-Bloch equations. It allows us to take into consideration the energy exchange between frequency components of radiation.

2. Mathematical model and results

We consider the medium of three-level atoms which has two lower levels and one excited level. In this system the quantum transitions are excited by the electromagnetic fields which have closed contour of the excitation. In this system there are two optical range transitions and one microwave range transition (Δ - scheme). Each transition is bound by the resonant electromagnetic field (figure 1 (a)). We suppose that the medium is optically thick (figure 1 (b)). It means that $n_a \sigma L > 1$, where n_a is the concentration of atoms, σ is the cross section of scattering of photons by an atom and L is the length of the medium (figure 1 (b)). Simultaneously the average distance between atoms is sufficiently large: $n_a \lambda^3 < 1$, where λ is the wavelength of the scattered radiation. The latter assumption allows us to ignore collective effects. The electric dipole transitions $|1\rangle \leftrightarrow |3\rangle$ and $|2\rangle \leftrightarrow |3\rangle$ interact with the fields of optical range with the Rabi frequencies Ω_1 and Ω_2. The magnetic dipole transition $|1\rangle \leftrightarrow |2\rangle$ interacts with the field of microwave range with the Rabi frequency *U*. Thus the three fields form a closed contour of excitation. It leads to dependence of optical properties of the medium on the relative phase $\Phi_0 = \varphi_1 - \varphi_2 - \varphi_3$ of fields at the entrance, where φ_j are the initial phases of each field (*j*=1,2,3).

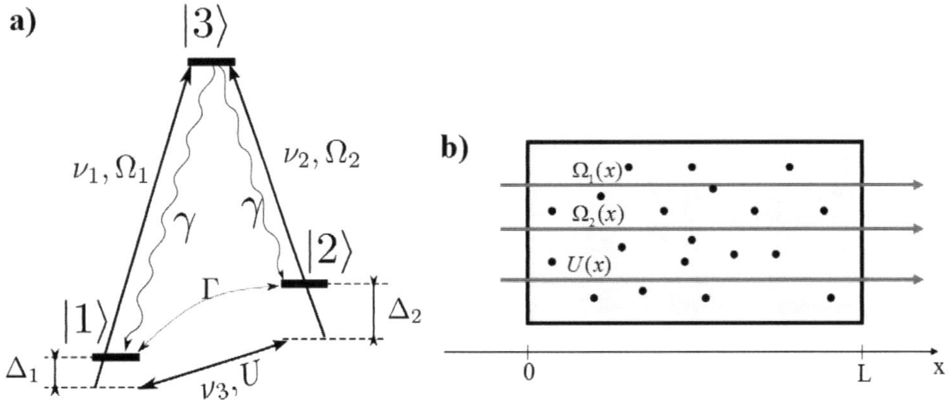

Figure 1. (a) – Level system and excitation scheme of atoms (Δ - scheme). Here ν_1 and ν_2 are the frequencies of radiation of optical range, ν_3 is the frequency of microwave field; Ω_1 и Ω_2 are the Rabi frequencies of radiation of optical range; U is the Rabi frequency of microwave field; γ is the rate of decay of excited level; Γ is the rate of decay of microwave coherence; Δ_1 and Δ_2 are the one-photon detunings. (b) – Propagation of the three-frequency radiation in the medium, which consists of atoms with concentration n_a non-interacting with each other.

The problem is described by the Maxwell-Bloch equations (1) for the slowly changing amplitudes of electric field $E_j^0(x,t)$ of waves and the atomic density matrix $\tilde{\rho}_{ik} = \rho_{ik} e^{i\nu_j t}$:

$$\begin{cases} \dfrac{\partial E_j^0(x,t)}{\partial x} + \dfrac{1}{c}\dfrac{\partial E_j^0(x,t)}{\partial t} = 4\pi P_j^0(x,t) k_j \\ \dfrac{\partial \tilde{\rho}_{ik}}{\partial t} = -\dfrac{i}{\hbar}\sum_l [H_{il}\tilde{\rho}_{lk} - \tilde{\rho}_{il} H_{lk}] + \sum_{l,m} \Gamma_{ik,lm}\tilde{\rho}_{lm} \end{cases}, \qquad (1)$$

where the index *j*=1,2 corresponds to each optical field. The wave vectors $\vec{k}_j = \vec{e}_x k_j$ are parallel to the axis *x*. *H* is the Hamiltonian and Γ is the relaxation matrix. $P_j^0(x,t)$ is the slowly changing amplitude of polarization of the medium. *c* is the speed of light, $\hbar$ is the Plank constant. We suppose that the microwave field *U* does not decay in the medium because the magnetic dipole moment of microwave transition $|1\rangle \leftrightarrow |2\rangle$ is 10^4 times less than electric dipole moments of optical transitions. Therefore, $U(x,t) = U(t)$. Below we consider the stationary case: the derivatives are equal to zero. Solution of the Maxwell-Bloch equations gives us spatial evolution of the optical fields $\Omega_j(x)$ and the slowly changing amplitudes of atomic density matrix $\rho_{ij}(x)$ in the medium. The values $\mathrm{Re}(\rho_{13})$ and $\mathrm{Re}(\rho_{23})$ are proportional to the resonant indexes of refractions of the corresponding transitions.

In figure 2 one can see the dependencies of index of refraction on x-coordinate for two sets of boundary conditions:

1 - $\Phi_0 = 0$ and $\Omega_1(0) \neq \Omega_2(0)$ (figure 2 (a,c));

2 - $\Omega_1(0) = \Omega_2(0)$ and $\Phi_0 = \pi / 4$ (figure 2 (b, d)).

In the case of different Rabi frequencies of the optical fields and zero relative phase the indexes of refraction of two optical transitions are not equal to zero and to each other. This leads to phase shifts of the fields along the medium and prevents transition to the state of electromagnetically induced transparency.

In the case of equal Rabi frequencies and nonzero relative phase the phase shifts arise due to nonzero relative phase which destroys the state of electromagnetically induced transparency. On the figure 2 (c,d) one can see spatial evolution of the relative phase. Also spatial oscillations of intensity take place due to energy exchange between the frequency components of radiation. Such oscillations induce a quasi-periodic spatial distribution of the index of refraction in the medium which is typical for photonic crystals (figure 2 (a,b)).

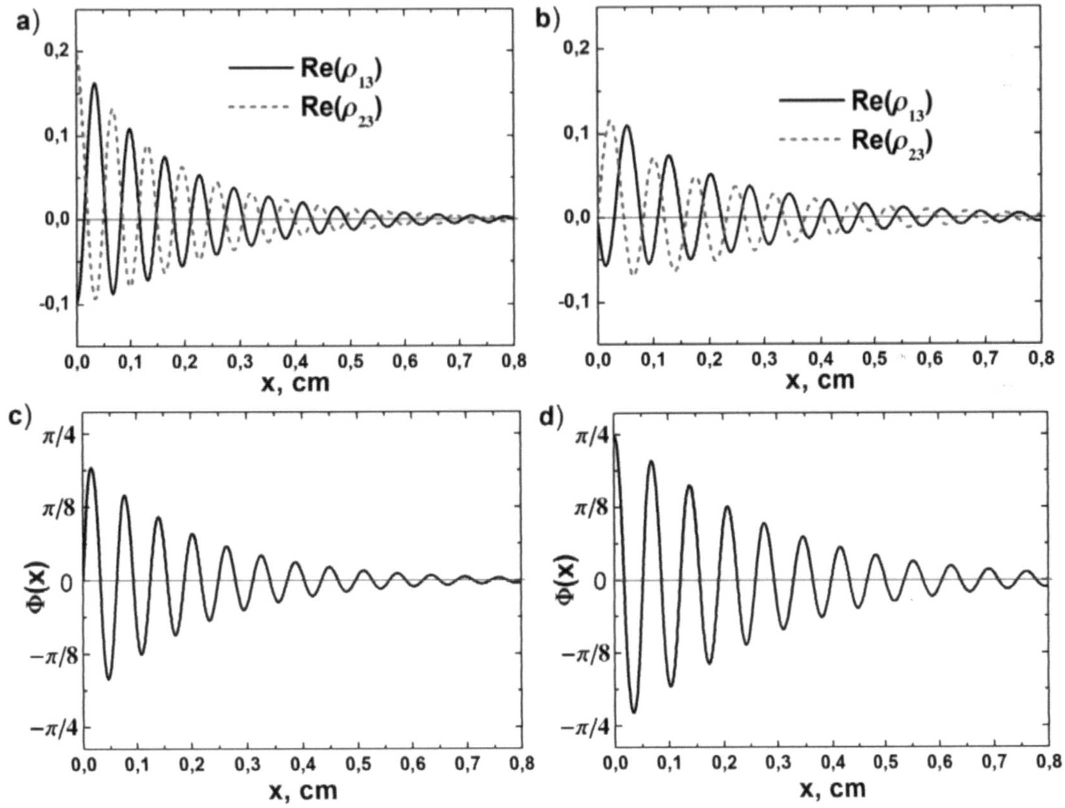

Figure 2. Dependencies of the real part of amplitude of atomic coherence of the transitions $|1\rangle \leftrightarrow |3\rangle$ and $|2\rangle \leftrightarrow |3\rangle$ and the relative phase Φ on the coordinate x for the boundary conditions, corresponding to the first way (a,c) and the second way (b,d) of excitation of the spatial oscillations. Parameters for (a,c): $\Phi_0 = 0$, $\Omega_1(0) = \gamma$, $\Omega_2(0) = 2\gamma$, $U = 3.5\gamma$, $n_a = 2*10^{11} cm^{-3}$; Parameters for (b,d): $\Phi_0 = \pi / 4$, $\Omega_1(0) = \Omega_2(0) = \gamma$, $U = 3.5\gamma$, $n_a = 2*10^{11} cm^{-3}$.

In figure 2 the envelope of oscillations of $\mathrm{Re}(\rho_{13})$ and $\mathrm{Re}(\rho_{23})$ decreases exponentially. If we enhance intensity of the optical fields the probability of stimulated emission will grow up and probability of spontaneous emission will fall down. It leads to reduction of radiation losses sideways therefore envelope of intensity of the optical fields and the real part of amplitude of atomic coherences falls down slower. For increasing the value of resonant index of refraction it is necessary to gain the high atomic concentration. For the value of atomic concentration $n_a \sim 10^{16}\ cm^{-3}$ the amplitude of refraction oscillations will be tens percent of non-resonant value [4,5]. Such concentrations can be achieved in rare earth crystals [3].

3. Conclusions

In this work we investigate interaction of the medium which consists of the three-level atoms with two electromagnetic fields of optical range and one microwave field. The fields form a closed contour of excitation (Δ - scheme). The calculation was made for the optically dense medium. It is found that two types of boundary conditions lead to pumping of energy between the optical fields, and index of refraction has the spatial quasi-periodic dependence. We analyzed such dependencies for different conditions of excitation and identified the boundary conditions which provide maximal amplitude of oscillations of the index of refraction. For enhancing of this amplitude it is necessary to increase the atomic concentration and intensity of the fields. The spatial quasi-periodic dependence of the index of refraction can be used for creation of the medium with controllable photonic band. Such schemes may be implemented in atomic gases as well as in solid state matrix.

Acknowledgments

This work was supported by Federal Special-Purpose Program "Science- and science-pedagogical personnel of innovation Russia on 2009-2013" (contract № 16.740.11.0586, grant №14.B37.21.1938), Russian President grant for young candidates of sciences (project MK-384.2013.2) and non-profit foundation "Dynasty".

Authors are grateful to Sokolov I.M. and Losev A.S. for useful discussions of this paper.

References

[1] Yablonovitch E 1987 *Phys. Rev. Lett.* **58** 2059
[2] John S 1987 *Phys. Rev. Lett.* **58** 2486
[3] Chris O'Brien and Kocharovskaya O 2011 *Phys. Rev. Lett.* **107** 137401
[4] Crenshaw M E, Bowden C M and Scully M O 2003 *J. Mod. Opt.* **50** 2551
[5] Rebane A K, Thiel C W, Mohan R K and Cone R L 2010 *Bull. Russ. Acad. of Sci. Phys.* **74** 891

Journal of Physics: Conference Series **478** (2013) 012009 doi:10.1088/1742-6596/478/1/012009

Effective defect mode suppression in a magnetophotonic crystals in the magnetic resonance region

Svetlana V. Eliseeva*, **Vladimir A. Ostatochnikov, and Dmitrij I. Sementsov**

Department of High Technology Physics and Engineering, Ulyanovsk State University, Lev Tolstoy 42, 432700, Ulyanovsk, Russian Federation

E-mail: *eliseeva-sv@yandex.ru

Abstract. The influence of magnetic field on the reflection spectra of one-dimensional magneto-photonic crystals of two types has been investigated. The first type is a periodically layered defect-free magnetic-dielectric structure, the second one is a magnetically active defect placed between dielectric photonic crystal mirrors. If the frequency of magnetic resonance is close to the central frequency of one of the photonic band gaps or to the frequency of the defect mode, it creates a significant reconstruction of the spectrum in the mentioned frequency domain. Namely, it can bring the suppression of the oscillation spectrum and the defect mode, which makes it possible to effectively control the spectrum of such a structure by an external magnetic field.

1. Introduction

One-dimensional photonic-crystal (PC) structures and layered periodic structures made of various metarials have gained much attention of researchers over the last years. Owing to periodic modulation of the refractive index, the photonic spectrum of these structures possesses the forbidden band gaps in which the incident radiation is reflected almost completely [1, 2, 3, 4]. This property is important for various applications of PC structures, in particular, for controlling the optical radiation in laser engineering and data transmission systems. These structures functionality can be significantly extended by controlling their spectral characteristics via the variation of the geometrical or physical parameters of the structure. In particular, the photon spectrum of the structure can be modified by introducing the active magnetic layers into the structure, as a result the spectrum can be controlled by the external magnetic field [5, 6, 7, 8]. A significant spectrum reconstruction can also be caused by a local breaking of the structures periodicity that can lead to a defective transmission miniband in the forbidden band gap [9, 10, 11, 12]. The structure, which has a defect layer placed between two Bragg photonic crystal mirrors (Fabry-Perrot microcavity), is of special interest. This layer in such a structure plays the role of an optical microcavity, where the light wave field can be localized, thus greatly enhancing many effects of the light-matter interaction [8, 9, 10]. In such a structure, this layer plays the role of the optical microcavity where the light wave field can be localized and various effects of interaction of radiation with media can be significantly increased [13, 14, 15]. An effective control over the photonic spectrum, by using the external magnetic field, is also possible in microcavity structures with a magnetic defect. The resonance response of the magnetic permeability of the defect to the high frequency field of the propagating wave in the region of magnetic resonance, can substantially modify the spectral line of the defect mode up to its complete suppression. In this work, we investigate the modification of reflection spectra of a magnetic-dielectric defect-free PC structure and microcavity structure with a magnetic defect in the external magnetic field in the domain of magnetic resonance, as well as the possibility of suppression of the oscillation spectrum and the defect mode of the spectra concerned.

2. Material parameters of layers and transfer matrices

Let us consider two types of one-dimensional magnetoactive PC structures. The first type is the finite defect-free periodically layered structure consisting of alternating magnetic and dielectric layers. The second structure type is a symmetrical magnetoactive microcavity, in which the magnetic layer is placed between the dielectric PC mirrors.

The structures of the first type include non-magnetic dielectric layers, which is characterized by the thickness L_d and scalar permittivity and permeability (DP and MP) ε_d and μ_d. Layers of two different dielectrics with thicknesses $L_{1,2}$, permittivities $\varepsilon_{1,2}$ and permeabilities $\mu_{1,2}$ are used in the structures of the second type. In the structures of both types the layers of uniformly magnetized magnetic are characterized by thickness L_m and are described by the scalar DP and tensor MP in the high frequency range:

$$\hat{\mu}_m = \begin{pmatrix} 1 & 0 & 0 \\ 0 & \mu & -i\mu_a \\ 0 & i\mu_a & \mu \end{pmatrix}. \tag{1}$$

We assume that the external static magnetic field H_0 is oriented along the OX-axis. Tensor components $\hat{\mu}_m$ are the following:

$$\mu = 1 + \frac{\omega_M(\omega_H + i\alpha\omega)}{(\omega_H + i\alpha\omega)^2 - \omega^2}, \quad \mu_a = \frac{\omega_M\omega}{(\omega_H + i\alpha\omega)^2 - \omega^2}, \tag{2}$$

where we introduce the parameters $\omega_H = \gamma H_0$, $\omega_M = 4\pi\gamma M_0$, M_0 is the saturation magnetization, γ is the gyromagnetic ratio, $\alpha = \Delta H/H_0$ is the parameter of the magnetic damping, ΔH - is the resonance line width [16].

We assume that the waves propagate in the structure along its periodicity axis (the OY axis), and the field H_0 is oriented perpendicular to this axis, i.e. lies in the layers plane. The solution of the Maxwells equations, taking into account the propagation direction and the orientation of the bias field, yields two eigenmodes of the structure, the TE and TM modes. Only TE wave with the components of wave field (E_x, H_y, H_z) is controlled by an external magnetic field. TM-type wave with (H_x, E_y, E_z) components practically does not react to changes in the external magnetic field.

Solutions of wave equations for each component in j-th layer can be represented as a superposition of the forward and backward waves:

$$F_{\alpha,j}(y) = F^1_{\alpha,j} \exp\left[i(\omega t - k_j y)\right] + F^2_{\alpha,j} \exp\left[i(\omega t + k_j y)\right], \tag{3}$$

where $k_j = k_0\sqrt{\varepsilon_j\mu_j}$ are the wave numbers in each of the layers, $k_0 = \omega/c$, c is the light speed in vacuum. For dielectric layers $\varepsilon_j = \varepsilon_d$, $\mu_j = \mu_d$. For the magnetic layers in the case of TM wave $\varepsilon_j = \varepsilon_m$, $\mu_j = 1$, and in the case of TE wave

$$\mu_j = \mu_\perp = \mu - \frac{\mu_a^2}{\mu} = \frac{(\omega_a + i\alpha\omega)^2 - \omega^2}{\omega_r^2 - \omega^2 + i\alpha\omega(\omega_a + \omega_H)}, \tag{4}$$

where we introduce the magnetic resonance frequency $\omega_r = \sqrt{\omega_H \omega_a}$ and antiresonance frequency $\omega_a = \omega_H + \omega_M$ [16]. The frequency and field dependences of $\mu_\perp(\omega, H_0)$ determine in many ways the characteristics of TE wave interaction with the magnetoactive PC structure.

Figure 1 shows the dependences of the real part of the effective permeability on the frequency for the field values $H_0 = (400; 600)\ Oe$ (the solid curves 1, 2), and the magnitude of the magnetizing field for the frequency values $\omega = (1.5; 2.0) \cdot 10^{10} s^{-1}$ (the dashed curves 3, 4). The calculations were made for a magnetic with parameters $4\pi M_0 = 1780$ G, $\alpha = 0.05$. It can be seen that each value corresponds to the magnetizing field value of resonance frequency, and each frequency corresponds to the resonance value of the magnetic field.

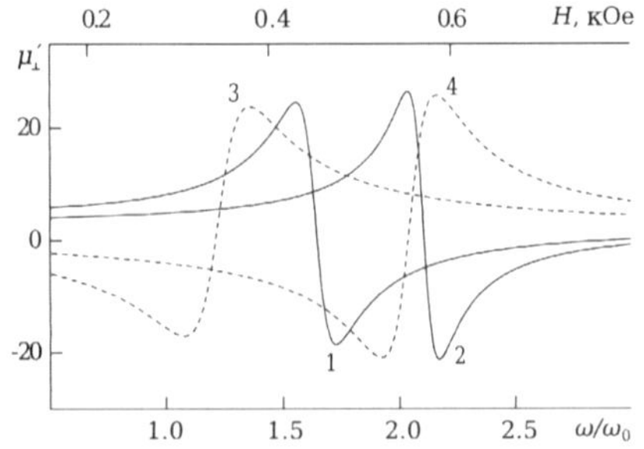

Figure 1. Frequency dependences of the real part of the effective magnetic permeability for the values of the field $H_0 = (0.4; 0.6)\ kOe$ (solid curves 1, 2) as well as the value of the magnetizing field for values of the frequencies $\omega = (1.5; 2.0) \cdot 10^{10}\ s^{-1}$ (dashed curves 3, 4); $\omega_0 = 10^{10} s^{-1}$.

The increase of the field leads to a shift of the resonance line to higher frequencies, and the increase in the frequency to the region of high fields. Let us consider the frequency region $\omega_r < \omega < \omega_a$, where the real part of the effective permittivity $\mu_\perp$ is negative (here we introduced the frequency antiresonance $\omega_a = \omega_H + \omega_M$). By changing the magnetic field, these features of the functions $\mu_\perp(\omega, H_0)$ allow to provide proximity magnetic resonance frequency or region of negative effective MP to the center frequency of the photonic band gap, or to the frequency of the defect mode. This, in turn, should lead to a substantial restructuring of the photon spectrum of a magnetophotonic crystal.

Using the boundary conditions for wave fields and the periodicity, we can get the connection of wave fields in planes separated by an arbitrary number of layers. In the case of defect-free periodic structure containing a finite number of periods, this relationship is a matrix $\hat{G} = (\hat{M})^n$, which is the n-th power of the matrix of one period. In its turn, the transfer matrix of one period is the product of the transfer matrices of the layers that make up the period, i.e. $\hat{M} = \hat{N}_1 \hat{N}_2$. In this case, the transfer matrices of each layer are as follows:

$$\hat{N}_j = \begin{pmatrix} C_j & \dfrac{k_0 \mu_j}{i k_j} S_j \\ -\dfrac{i k_j}{k_0 \mu_j} S_j & C_j \end{pmatrix}, \tag{5}$$

where we introduced the notation $C_j = \cos(k_j L_j)$, $S_j = \sin(k_j L_j)$. Using (5) the transfer matrix of one period has the following matrix elements:

$$\begin{aligned} M_{11} &= C_1 C_2 - \frac{\mu_1 k_2}{\mu_2 k_1} S_1 S_2, \quad M_{12} = -\frac{k_0 \mu_2}{i k_2} C_1 S_2 - \frac{k_0 \mu_1}{i k_1} S_1 C_2, \\ M_{21} &= \frac{i k_2}{k_0 \mu_2} C_1 S_2 + \frac{i k_1}{k_0 \mu_1} S_1 C_2, \quad M_{22} = C_1 C_2 - \frac{\mu_2 k_1}{\mu_1 k_2} S_1 S_2, \end{aligned} \tag{6}$$

The symmetric PC resonator structure is intended to include a layer of uniformly magnetized magnetic between the two lateral dielectric photonic-mirrors. We assume that each of the side PC-mirrors consists of the same number of structure periods, and the period is composed of two layers of non-absorbing dielectrics. The inversion means changing the order of the layers of one part of the structure against the other one. In terms of defects such structure contains a double defect - the inversion layer and the introduction layer [12]. Inverted period corresponds to the transfer matrix $\hat{\overline{M}} = \hat{N}_2\hat{N}_1$, whose matrix elements are related to the elements of the matrix of the normal period by the ratio $\hat{\overline{M}}_{\alpha,\beta} = \hat{M}_{3-\alpha,3-\beta}$, where $\alpha, \beta = 1, 2$. Transfer matrices of two types of considered resonator structures can be written as $\hat{G} = (\hat{M})^5\hat{N}_m(\hat{\overline{M}})^5$, $\hat{G} = (\hat{\overline{M}})^5\hat{N}_m(\hat{M})^5$.

Energy reflection and transmission coefficients for the investigated PC structures can be represented as follows:

$$R = \Big(\frac{G_{11} + G_{12} - G_{21} - G_{22}}{G_{11} + G_{12} + G_{21} + G_{22}}\Big)^2, \quad T = \frac{4}{(G_{11} + G_{12} + G_{21} + G_{22})^2}, \tag{7}$$

where $G_{\alpha,\beta}$- are the matrix elements of the transfer matrix structure. These coefficients satisfy the law of conservation of energy $R + T + A = 1$, where the absorption coefficient A determines the amount passed into heat energy.

3. Defect-free PC-structure magnetic-dielectric.

Based on the above relations, we performed an analysis of the reflection and transmission spectra of a defect-free periodic structure containing a finite number of periods, and investigated their modification under the influence of an external magnetic field. The transfer matrix of a structure consisting of n periods is the n - th power of the transfer matrix of one period, i.e., $\hat{G} = (\hat{M})^n$. To simulate reflective and selective properties of a PC structure we use the following values of permittivity $\varepsilon_d = 10$ and permeability $\mu_d = 1$ of layers (material MCT10) and $\varepsilon_m = 15.1$, function $\mu_\perp(\omega, H_0)$ is the same as in the figure 1 (doped yttrium iron garnet 10CH6B) [17].

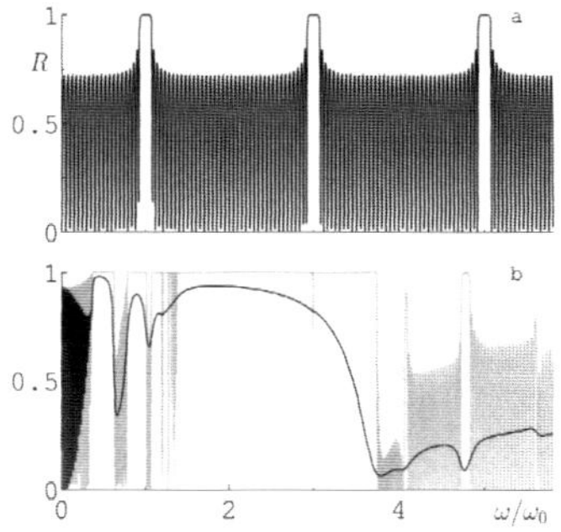

Figure 2. The reflection spectra of structure $(M)^{20}$ with thicknesses of the layers in $L_d = 1.49$ cm and $L_m = 1.21$ cm with and without application of an external field $H_0 = 0.3$ kOe (a and b); $\omega_0 = 10^{10}$ s^{-1}.

Figure 2 shows the reflection spectra of the structure $(M)^{20}$ with layer thicknesses $L_d = 1.49$ cm and $L_m = 1.21$ cm. The optical thickness of the dielectric layers $L_{0d} = L_d\sqrt{\varepsilon_d\mu_d} \simeq 4.71$ cm and the thickness of magnetic material layers has been chosen to ensure their optical depth $L_{0m} = L_m\sqrt{\varepsilon_m\mu_m} = L_{0d}$ at value $\mu_m = 1$. Figure 2a shows the reflection spectrum for TM wave, which effectively has no frequency dispersion of values $\mu_j = 1$ in the entire range. This reflection spectrum has all the features specific to a one-dimensional dielectric PC structure [3, 4]. Figure 2b shows the reflection spectrum of the TE wave which is significantly affected by the frequency dependence of the effective magnetic permeability $\mu_\perp$. Spectra correspond to the external field value $H_0 = 300$ Oe, as well as to the presence of magnetic damping ($\alpha = 0.05$, black line) and to its absence ($\alpha = 0$, gray line). It can be seen that the presence of the damping reduces the maximum coefficient R values, as compared with its value in the absence of damping. In both cases, the comparison of the TE wave spectra with TM wave spectrum denotes their substantial transformation.

4. Inverted structure with a magnetic defect

Now let us consider the symmetric PC resonator structure. Each of the side PC dielectric mirrors consists of five periods, each period consists of two layers of non-absorbing dielectric with real constants $\varepsilon_{1,2}, \mu_{1,2} = 1$, and thicknesses $L_{1,2}$. The thickness of the cavity layer of uniformly magnetized magnetic is L_m. Side mirrors are inverted from each other.

To simulate high-frequency properties of this structure we use the following values of permittivity layers of PC-mirrors: $\varepsilon_1 = 10$, $\varepsilon_2 = 25$ (materials MCT10 and MCT25) [17]. We choose doped yttrium iron garnet 10CH6B, whose characteristics are already described above, as the defect layer material. Thickness of layers were chosen to ensure the same optical thickness of the dielectric layers of PC mirrors, i.e. $L_1\sqrt{\varepsilon_1} = L_2\sqrt{\varepsilon_2} = L_0$, and the optical thickness of the magnetic layer was taken equal to double optical thickness of the dielectric layers, i.e. $\sqrt{\varepsilon_m < \mu'_\perp >} L_m = 2L_0$, where $< \mu'_\perp >= 2.57$ (average permeability in the field of magnetic resonance). The optical thickness $L_0 \simeq 24.4\ mm$. For the considered structures the position and width of the spectral line of the defect mode depend stronger on the sequence of layers in the PC-mirrors as well as on the value of $< \mu'_\perp >$.

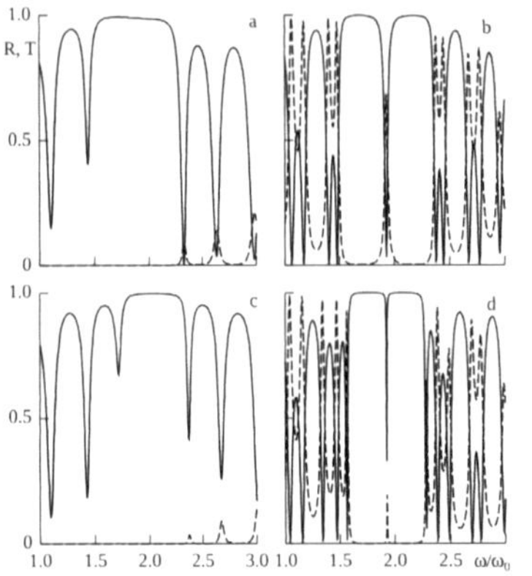

Figure 3. The reflection and transmission spectra (solid and dashed curves) for the structure $(M)^5N_m(\overline{M})^5$ (a, b) and $(\overline{M})^5N_m(M)^5$ (c, d) prepared for field values of $H_0 = 0.2\ kOe$ (a, c) and $H_0 = 1.5\ kOe$ (b, d); $\omega_0 = 10^{10}\ s^{-1}$.

Figure 3 shows the reflection and transmission spectra (solid and dashed lines) for the structures $(M)^5N_m(\overline{M})^5$ and $(\overline{M})^5N_m(M)^5$, obtained for different values of the field $H_0 = 0.2\ kOe$ (a, c) and $H_0 = 1.5\ kOe$ (b, d), $< \mu'_\perp >= 2.57$. At the value of the field $H_0 = 0.2\ kOe$ the resonant frequency $\omega_r = 1.1 \cdot 10^{10}\ s^{-1}$ and the spectra correspond to the case when the photonic band gap coincides with the domain of "metallic reflection", for which the values are negative. At $H_0 = 1.5\ kOe$, $\omega_r = 3.9 \cdot 10^{10}\ s^{-1}$ the situation is shown which corresponds to the equation $\mu'_\perp =< \mu'_\perp >$, where the defect mode is in the center of the band gap. Further increase of the magnetizing field leads to a shift of the resonant frequency to the optical frequency range, when the value $\mu'_\perp$ becomes smaller $< \mu'_\perp >$ and the defect mode is shifted to high frequancy edge of the band gap.

Figure 4 shows the reflectance spectra of the structure $(\overline{M})^5N_m(M)^5$ with thicknesses of layers $L_1 = 7.72$ mm, $L_2 = 4.88$ mm with the same optical thickness $L_0 = 24.4$ mm (at the frequency $\omega_0 = 10^{10}\ s^{-1}$) and with defect thicknesses $L_m = (0.001; 0.01; 2.0)L_0$ and $L_m = (0.001; 0.01; 2.0)\lambda_r$ (a, b; solid, dashed, dotted lines). The spectra were obtained at $H_0 = 0.5$ kOe, the resonant frequency $\omega_r = 1.88 \cdot 10^{10}\ s^{-1}$ and $\lambda_r = 2\pi c/\omega_r \simeq 10\ cm$. It reveals the substantial dependence of the spectrum character on the thickness of the defect layer near the defect mode. The thicker it is, the stronger the effect of the magnetic resonance spectrum becomes. If the thickness of the defect layer is the same or greater than the optical thickness, the defect mode is weak or completely suppressed. With a significant decrease in the thickness of the defect layer the defect mode is manifested associated with a defect inversion in the structure.

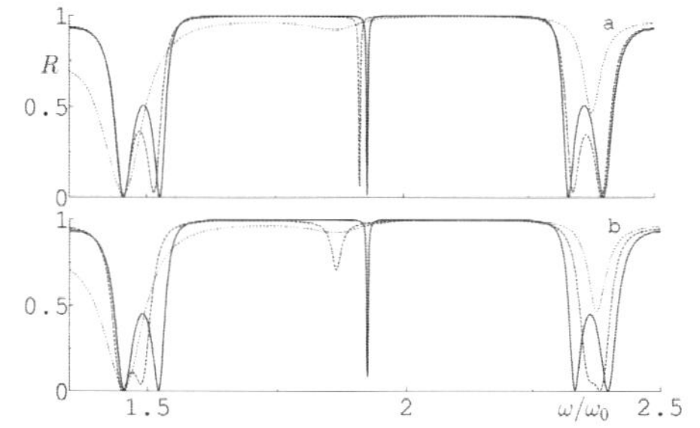

Figure 4. The reflection spectra of the structure $(\overline{M})^5 N_m (M)^5$ with thicknesses of the layers in $L_1 = 7.72\ mm$ and $L_2 = 4.88\ mm$; $L_m = (0.001; 0.01; 2)L_0$ and $L_m = (0.001; 0.01; 2)\lambda_r$ (a, b; solid, dashed, dotted lines); with an external field $H_0 = 0.5\ kOe$; $\omega_0 = 10^{10}\ s^{-1}$.

5. Conclusion

The analysis carried out in the work suggests the possibility of effective control over reflectivity and transmissivity of a magnetoactive PC structure in the region of magnetic resonance by an external magnetic field. For TE waves we demonstrated almost complete suppression of the oscillations of the reflection coefficient of a defect-free PC and of the defect mode in the case of a defect structure at the coincidence of the magnetic resonance with the frequency doimain of the photonic band gap. In the case of propagation of TM wave in these structures, there is no resonance in the magnetic layers, so the management of its wave characteristics by an external magnetic field is not possible. Thus, the magnetic sensitivity of the defect mode in the case of TE polarization is nonmagnetic sensitive in the case of TM polarization, i.e. the defect mode in the above structure is polarization-sensitive. The detected effects can be used as the basis for the creation of such high-frequency radiation control units, as modulators, filters and switches.

Acknowledgments

This work was supported by the Russian Foundation for Basic Research (state contract No.13-02-97026) and by the Ministry of Education and Science of the Russian Federation

References

[1] Inoue K and Ohtaka K 2004 *Photonic crystals: physics, fabrication and applications* vol 94 (Springer)
[2] Sakoda K 2005 *Optical properties of photonic crystals* (Springer Verlag)
[3] Johnson S and Joannopoulos J 2002 *Photonic crystals: the road from theory to practice* (Kluwer Academic Publishers) ISBN 9780792376095
[4] Eliseeva S V and Sementsov D I 2010 *Optics and Spectroscopy* **109** 729 ISSN 1562-6911
[5] Gulyaev Y and Nikitov S 2003 *Radiotechnics* **26**
[6] Nikitov S and Tailhades P 2001 *Optics Communications* **199** 389–397 ISSN 0030-4018
[7] Inoue M, Fujii T and Abe M 1998 *Journal of Applied Physics* **83** 6768
[8] Figotin A and Vitebskiy I 2003 *Physical Review B* **67** 165210 ISSN 1550-235X
[9] Steel M, Levy M and Osgood Jr R 2000 *Photonics Technology Letters, IEEE* **12** 1171–1173
[10] Eliseeva S and Sementsov D 2011 *Journal of Experimental and Theoretical Physics* **112** 199–203
[11] Lyubchanskii I, Dadoenkova N, Lyubchanskii M, Shapovalov E and Rasing T 2003 *Journal of Physics D: Applied Physics* **36** R277
[12] Eliseeva S, Ostatochnikov V and Sementsov D 2013 *Physics of the Solid State* **55** 73–76
[13] Erokhin S, Vinogradov A, Granovsky A and Inoue M 2007 *Physics of the Solid State* **49** 497–499 ISSN 1063-7834
[14] Eliseeva S and Sementsov D 2012 *Physics of the Solid State* **54** 1981–1987
[15] Born M and Wolf E 1999 *Principles of optics* 7th ed (Cambridge University Press) ISBN 0521642221
[16] Gurevich A G and Tybulewicz A 1963 *Ferrites at microwave frequencies* (Consultants Bureau New York)
[17] *http://www.magneton.ru* Parameter tables made of Microwave ferrites and dielectrics. Magneton J.S.Co. St–Petersburg, Russia.

Optical vibrations in a closed chain of coupled waveguides or cavities

A P Sukhorukov, I V Savochkin

Lomonosov Moscow State University, Moscow, Russia

E-mail: apsmsu@gmail.com

Abstract. We study the dynamics of light field in tunnel-coupled oscillators of two types such as cavities and waveguides. The introduction of linear and nonlinear defects leads to localization and screening fields. Rotating modes and discrete vortices are considered by analytical solutions and numerical simulation.

1. Introduction

Systems of tunnel-coupled nonlinear optical waveguides and cavities are of interest in connection with the potential applications such as optical couplers, elements of optical logical devices, etc. [1, 2, 3, 4, 5, 6]. Therefore detailed understanding of the basic characteristics of such structures is a very important component for new effects research in nonlinear and coherent optics.

2. Basic equations of the model

We study the spatial dynamics of the three oscillators using three waveguides model. Obtained results are easily transferred to the description of the temporal dynamics of coupled cavities. In general, one of the oscillators has a different eigenfrequency or refractive index. External radiation is introduced into one of the oscillators. Then the light propagation dependence on the basic parameters of the system with and without the defectis is considered. The effects of radiation localization and waveguide screening are investigated in detail.

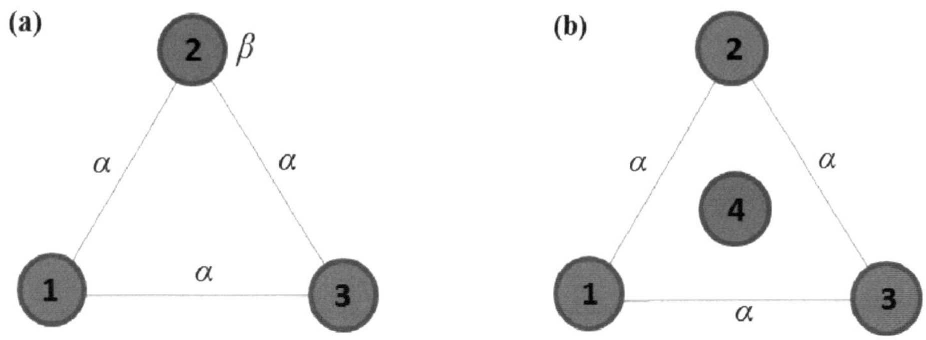

Figure 1. Cross-section of the system of three (a) and four (b) tunnel-coupled optical waveguides.

We consider discrete systems shown in figure 1a. The second waveguide is defective.

The system under consideration consists of three tunnel-coupled waveguides, i.e. light beam energy will penetrate from one of the waveguides into the neighboring, thus exciting them. Coupling coefficient describes the rate of energy transfer between the waveguides. It depends on the distance between the waveguides and dielectric permittivity of substrate.

Amplitudes of the light in the waveguides can be found from these equations:

$$\begin{aligned} \frac{\partial C_1}{\partial z} &= -i\alpha(C_2 + C_3) \\ \frac{\partial C_2}{\partial z} &= -i\alpha(C_1 + C_3) + i\beta|C_2|^2 C_2 \\ \frac{\partial C_3}{\partial z} &= -i\alpha(C_2 + C_1) \end{aligned} \tag{1}$$

where $C_j = |C_j|e^{i\phi_j}$ is the complex amplitude of the wave in the j-th waveguide, β is the coefficient of nonlinearity. In the theory of coupled cavities spatial coordinate x is replaced by the time coordinate t. From the system (1) we can get the following equations describing the variation of energy and phase of the light wave in the waveguides:

$$\begin{aligned} \frac{d|C_1|^2}{dz} &= -2\alpha|C_1||C_2|\sin(\phi_2-\phi_1) - 2\alpha|C_1||C_3|\sin(\phi_3-\phi_1) \\ \frac{d|C_2|^2}{dz} &= -2\alpha|C_2||C_1|\sin(\phi_1-\phi_2) - 2\alpha|C_2||C_3|\sin(\phi_3-\phi_2) \\ \frac{d|C_3|^2}{dz} &= -2\alpha|C_2||C_3|\sin(\phi_2-\phi_3) - 2\alpha|C_1||C_3|\sin(\phi_1-\phi_3) \end{aligned} \tag{2}$$

$$\begin{aligned} \frac{d\phi_1}{dz} &= -\alpha\left(\frac{|C_2|}{|C_1|}\cos(\phi_2-\phi_1) + \frac{|C_3|}{|C_1|}\cos(\phi_3-\phi_1)\right) \\ \frac{d\phi_2}{dz} &= -\alpha\left(\frac{|C_1|}{|C_2|}\cos(\phi_1-\phi_2) + \frac{|C_3|}{|C_2|}\cos(\phi_3-\phi_2)\right) + \beta|C_2|^2 \\ \frac{d\phi_3}{dz} &= -\alpha\left(\frac{|C_1|}{|C_3|}\cos(\phi_1-\phi_3) + \frac{|C_2|}{|C_3|}\cos(\phi_2-\phi_3)\right) \end{aligned} \tag{3}$$

The analysis of the equations (2) and (3) shows that the direction of energy transfer between the waveguides is determined by the phase difference of the light wave in the adjacent waveguides.

For $\beta = 0$ equations (1) can be solved analytically:

$$\begin{cases} C_1 = \dfrac{E_1+E_2+E_3}{3}e^{i2\alpha z} + \dfrac{2E_1-E_2-E_3}{3}e^{-i\alpha z} \\ C_2 = \dfrac{E_1+E_2+E_3}{3}e^{i2\alpha z} + \dfrac{2E_2-E_1-E_3}{3}e^{-i\alpha z} \\ C_3 = \dfrac{E_1+E_2+E_3}{3}e^{i2\alpha z} + \dfrac{2E_3-E_2-E_1}{3}e^{-i\alpha z} \end{cases} \tag{4}$$

where E_j is the initial amplitude of the wave in the j-th waveguide.

There is one important effect in the linear identical waveguides. At a certain ratio of initial phases (in our case $\phi_1 = 0$, $\phi_2 = \dfrac{2\pi}{3}$, $\phi_3 = \dfrac{4\pi}{3}$) and equal amplitudes of the incoming radiation the cyclical energy transfer from one waveguide to another occurs, so-called rotating mode. The essence of this phenomenon is that the amplitude of the radiation in the waveguides remains constant, but nevertheless energy is transferred between the waveguides due to fixed phase mismatch. How much energy is transferred per time unit from the first to the second waveguide the same is pumped from the third to the first waveguide, etc. So if we put fourth waveguide in the centre of our system (figure 1b) the energy will not penetrate into this waveguide. Therefore if the initial light amplitude at the input of the central waveguide is equal to zero, it will be zero along the whole waveguide. This effect can be called «discrete optical vortex». If the chain contains 6 oscillators, then it is possible to excite vortex with double charge: $\varphi_m = (m-1)\pi / 6$. Discrete double-charge vortices were thoroughly studied in [6].

In the general case it is possible to solve equations (1) only numerically, but for linear defect (i.e. one of waveguides has different refractive index) they transform to the equations (5), which can be solved and analyzed analytically:

$$\begin{aligned} &\frac{\partial C_1}{\partial z} = -i\alpha(C_2 + C_3) \\ &\frac{\partial C_2}{\partial z} = -i\alpha(C_1 + C_3) + i\Delta k\, C_2 \quad \frac{\partial C_3}{\partial z} = -i\alpha(C_2 + C_1) \end{aligned} \tag{5}$$

where k_j is the wave number for j-th waveguide, $k_1 = k_3$, and $\Delta k = k_2 - k_1$ – wave detuning. Solution:

$$\begin{aligned} C_1 &= \frac{E_1 - E_3}{2} e^{i\alpha z} + \frac{1}{2}(E_1 + E_3) e^{-i\left(\frac{\alpha - \Delta k + s}{2}\right)z} \\ C_2 &= -\frac{1}{4}(E_1 + 4E_2 + E_3) e^{-i\left(\frac{\alpha - \Delta k - s}{2}\right)z} + \frac{1}{4}(E_1 + E_3) e^{-i\left(\frac{\alpha - \Delta k + s}{2}\right)z} \\ C_3 &= -\frac{E_1 - E_3}{2} e^{i\alpha z} + \frac{1}{2}(E_1 + E_3) e^{-i\left(\frac{\alpha - \Delta k + s}{2}\right)z} \end{aligned} \tag{6}$$

From equations (6) we can see that localization occurs if the initial amplitude E_2 in the defective waveguide is much higher than the amplitudes in other waveguides, and conversely, if the initial amplitude E_2 is much lower than E_1 or E_3 we can observe the screening effect. As shown by the simulation results (see section 3) these conditions are also valid for nonlinear defect.

All the above considerations can be applied to the closed chain of coupled oscillators or cavities. Parameters will have different meaning, but the solution remains the same.

3. The simulation results

Curves on the following graphs, describing the dependence of light intensity in the waveguides on the distance, are marked by the number of the relating waveguide.

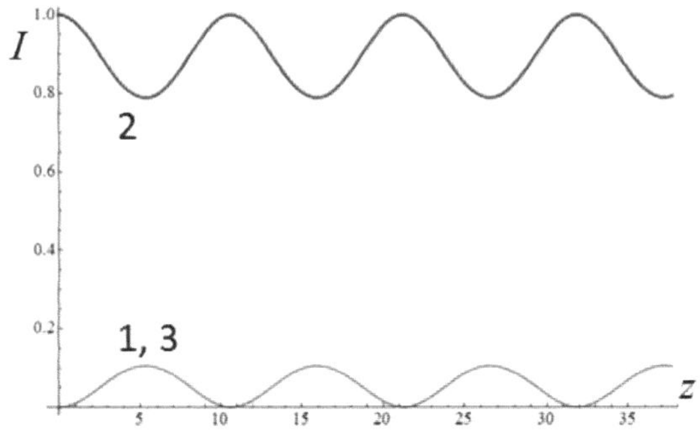

Figure 2. Localization of the radiation in the defective waveguide. $\alpha = 1$, $\beta = 0.5$.

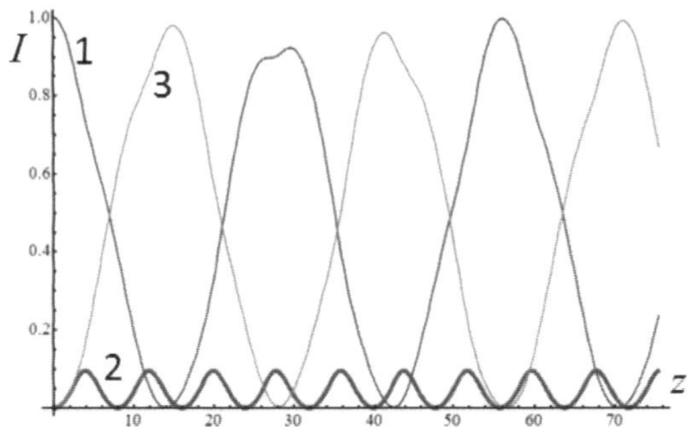

Figure 3. Defective waveguide screening. $\alpha = 1$, $\beta = 10$.

In this paper we show that almost all energy of the system can be concentrated either in a single waveguide (figure 2), or in two (figure 3), without going into the third, or in all three uniformly, cyclically flowing from one waveguide to another. Our system is closed, i.e. after supplying the initial radiation into the system its energy remains constant, but can be freely redistributed within the system, except cases in which it is not allowed by the physics of the process.

By analyzing numerical simulations of light propagation in waveguides with different parameters we have shown that after the excitation of the defect waveguide the radiation can be localized in it (figure 2), and the degree of localization increases with increasing nonlinearity coefficient (figure 4a) or input light intensity. If we excite not the defective waveguide, but any other, then the defective waveguide will be screened, i.e. the radiation will redistribute between two linear waveguides and almost will not pass through defective one. The screening level is also increased with increasing nonlinearity coefficient (figure4b).

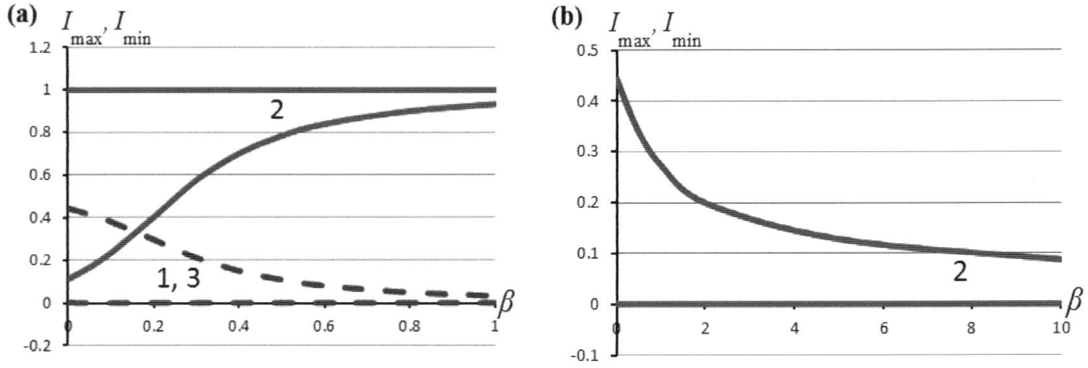

Figure 4. Dependence of the minimum and maximum of light intensity in the waveguides on the nonlinearity coefficient. Solid lines correspond to the defective waveguide. a) The localization effect. b) The screening effect.

4. Conclusions

We studied the localization of waves in nonlinear defective waveguide. It was shown that the degree of localization increases with increasing nonlinearity coefficient or intensity of input emission in the defective waveguide. An effect of defective waveguide screening from neighboring zero-defect waveguides was found. Conditions for excitation of discrete optical vortex in the coupled waveguides were found.

References

[1] Jensen S M 1982 *IEEE J. Quant. Electron* **18** 1580–83
[2] Schmidt-Hattenberger S, Trutschel U and Lederer F 1992 *Opt. Lett.* **24** 691–701
[3] Sukhorukov A A and Kivshar Yu S 2001 *Phys. Rev. Lett.* **87** 083901
[4] Fedele F, Yang J and Chen Z 2005 *Opt. Lett.* **30** 1506–08
[5] Moison J M, Belabas N, Minot C and Levenson J A 2009 *Opt. Lett.* **34** 2462
[6] Terhalle B, Richter T, Law K J H, Göries D, Rose P, Alexander T J, Kevrekidis P G, Desyatnikov A S, Krolikowski W, Kaiser F, Denz C, and Kivshar Yu S 2009 *Phys. Rev. A***79** 043821

Derivation of kinetic equations from non-Wiener stochastic differential equations

A M Basharov

Moscow Institute of Physics and Technology, 141700 Dolgoprudny, Russia
National Research Centre «Kurchatov Institute», 123182 Moscow, Russia

E-mail: basharov@gmail.com

Abstract. Kinetic differential-difference equations containing terms with fractional derivatives and describing α-stable Levy processes with $0 < \alpha < 1$ have been derived in a unified manner in terms of one-dimensional stochastic differential equations controlled merely by the Poisson processes.

1. Introduction

Stochastic differential equations (SDE) provide a straightforward and efficient technique of deriving kinetic equations to describe the dynamics of an open system both for classic cases [1] and quantum ones [2,3]. Nevertheless, before works [4-7] only Wiener-type SDEs were used to derive kinetic equations for both classic and quantum cases. In [4-7] density matrix equations for several models of quantum open systems were derived which allowed predicting and describing such effects as stabilization of excited quantum states of atomic ensembles and photon systems on the basis of quantum non-Wiener-type SDEs.

In classic physics, anomalous diffusion [8,9], single molecule radiation in disordered media [10] and other processes with exceptional (non-Gaussian) statistics and kinetics related to the Levy stochastic processes are in the front line of research. However, kinetic equations for such systems were derived not from SDEs with a noise source but by means of other indirect methods. For example, in work [11] the Chandrasekhar method [12] was used. In the meantime, it is common knowledge that both Wiener - and Levy processes can be defined on the basis of Poisson processes [13]. The Poisson–controlled SDEs appear to be non–Wiener type SDEs. The Poisson process with the intensity λ follows simple and intuitively obvious incremental algebra [14]:

$$dN(t)dN(t) = dN(t)\,,\ dN(t)dt = dtdt = 0\,,\ < dN(t) >= \lambda dt \tag{1}$$

which should seemingly simplify the derivation of not only kinetic equations but also simplify general theoretical analysis. Though until recently such an approach had not been available. In work [15] it was shown as a simple example that kinetic equations for the Levy processes can also be directly derived from SDEs.

2. The simplest kinetic equation derivation from SDE

The simplest classic stochastic process controlled by the Poisson process is an electric current in an electron tube. The appropriate SDE takes the following form [1]:

$$dX(t) = -\gamma X(t)dt + \beta dN(t)\,,\ -\gamma q_0 + \beta = 0\,, \tag{2}$$

where the value q_0 is the elementary charge. Hence it follows that a kinetic equation can be easily derived with the help of the method analogous to derivation [1] of the Fokker-Planck equation from the Wiener SDE when considering the increment $df(X(t)) = f(X(t)+dX(t)) - f(X(t))$ for any continuously differentiable function $f(x)$:

$$df(X(t)) = \tfrac{1}{1!}f'(X(t))dX(t) + \tfrac{1}{2!}f''(X(t))[dX(t)]^2 + ... = f'(X(t))[-\gamma X(t)dt + \beta dN(t)] +$$
$$+ \tfrac{1}{2!}f''(X(t))[dX(t)]^2 + ... = -f'(X(t))\gamma X(t)dt + \{f(X(t)+\beta) - f(X(t))\}dN(t)\,.$$

Upon averaging $< f(N(t)) >= \int f(x)p(x,t)dx$ over probability density function $p(x,t)$ we have

$$\int f(x)\tfrac{\partial}{\partial t}p(x,t)dx = -\int f'(x)\gamma x p(x,t)dx + \lambda\int[f(x+\beta) - f(x)]p(x,t)dx\,,$$

Whence the differential-difference kinetic equation follows

$$\tfrac{\partial}{\partial t}p(x,t) = \tfrac{\partial}{\partial x}[\gamma x p(i,t)] + \lambda[p(x-\beta,t) - p(x,t)]\,. \tag{3}$$

3. Transition from microscopic to macroscopic description in terms of SDE

Equations (2) and (3) describe phenomena at microscopic level. In order to have a description at macroscopic level, it is necessary to renormalize initial equations [16,17]. For this purpose let us separate out the compensated Poisson process $N_c(t) = N(t) - \lambda t$, $< N_c(t) >= 0$ in equation (2) and write it over in the form

$$dX(t) = (-\gamma X(t) + \lambda\beta)dt + \beta dN_c(t)\,.$$

Then renormalization involves substitutions of $X(t)$ and $N_c(t)$ for $\tilde{X}(t) = X(st)/\phi(s)$ and $\tilde{N}_c(t) = N_c(st)/\phi(s)$ where the scaling function $\phi(s) \to \infty$ at $s \to \infty$:

$$d\tilde{X}(t) = (-\gamma s\tilde{X}(t) + \frac{\lambda s\beta}{\phi(s)})dt + \beta d\tilde{N}_c(t)\,.$$

As a result, the increment

$$df(\tilde{X}(t)) = f'(\tilde{X}(t))[(-\gamma s\tilde{X}(t) + \frac{\lambda s\beta}{\phi(s)})dt + \beta d\tilde{N}_c(t)] + \tfrac{1}{2!}\beta^2 f''(\tilde{X}(t))[d\tilde{N}_c(t)]^2 + ...$$

and its average contains terms ($n \geq 2$) except for the time increment

$$d\tilde{N}_c(t)\,,\ < d\tilde{N}_c(t) >= 0\,,\ [d\tilde{N}_c(t)]^n = [\phi(s)]^{-n}dN(st)\,,\ < [d\tilde{N}_c(t)]^n >= s[\phi(s)]^{-n}\lambda dt\,.$$

If we set $s = \phi(s) = \lambda >> \gamma$ (for dimensionless quantities), then all terms with $n > 2$ within the average increment equation $< df(\tilde{X}(t)) >$ can be neglected, thus we derive a conventional Fokker-Planck equation

$$\tfrac{\partial}{\partial t}p(x,t) = \tfrac{\partial}{\partial x}[\tilde{\gamma}(x - I)p(x,t)] + \tfrac{1}{2}\beta^2\tfrac{\partial^2}{\partial x^2}p(x,t)\,, \tag{4}$$

where a macroscopic quantity – average current magnitude $I = \beta/\gamma$ – was introduced, along with renormalized rate $\tilde{\gamma} = \lambda\gamma$.

A scaled compensated Poisson process $\tilde{N}_c(t)$ at $s = \lambda \to \infty$ appears as a conventional Wiener process $W(t)$ with increment algebra [1]

$$dW(t)dW(t) = dt\,,\; dW(t)dt = dtdt = 0\,,\; < dW(t) >= 0\,. \tag{5}$$

Note that the relation of the scaled compensated Poisson process to the Wiener process is generally established by means of the characteristic function. In a similar way, one can consider rescaling of compensated Poisson process only.

Terms of SDE containing dt, $dW(t)$ and $dN(t)$ are conventionally named as drift, diffuse and jump.

4. The derivation of general kinetic equation from SDE

Let us consider the simplest general SDE with drift, diffuse and jump terms:

$$dX(t) = a(X,t)dt + b(X,t)dW(t) + \beta dJ(t)\,,\; \beta > 0\,, \tag{6}$$

where a jump term is represented as the increment of composite Poisson process $J(t) = \sum_{\delta} \delta N(\delta,t)$.

Poisson processes $N(\delta,t)$ with various δ are independent and different by intensity $\lambda(\delta)$ for different values of jump δ. Analogously, all Poisson processes included into $J(t)$ are statistically independent of the Wiener process, $dW(t)dN(\delta,t') = 0$.

Pursuing the described conventional scheme we have the increment

$$df(X) = \sum_{k=1}^{\infty} \frac{1}{k!} \frac{\partial^k f(X)}{\partial X^k} (a(X,t)dt + b(X,t)dW(t) + \beta dJ(t))^k =$$

$$= \left(\frac{\partial f(X)}{\partial X} a(X,t) + \frac{1}{2} \frac{\partial^2 f(X)}{\partial X^2} b^2(X,t) \right) dt + \frac{\partial f(X)}{\partial X} b(X,t)dW(t) + \sum_{\delta} \{ f(X + \beta\delta)) - f(X) \} dN(\delta,t).$$

Upon averaging with regard to the increment algebra drift and diffuse terms lead to a conventional Fokker-Planck equation. Transformation of a jump term depends on the function $\lambda(\delta)$.

If we take the Levy measure [13,14] in place of $\lambda(\delta)$

$$\lambda(\delta) = \begin{cases} a_L \delta^{-(\alpha+1)}, & \delta > 0; \\ b_L |\delta|^{-(\alpha+1)}, & \delta < 0, \end{cases} \quad a_L > 0,\; b_L > 0\,, \tag{7}$$

then the composite Poisson process will describe α-stable Levy process. In addition,

$$< \sum_{\delta>0} \{ f(X + \beta\delta) - f(X) \} dN(\delta,t) >= \int p(x,t) \int_0^{\infty} a_L \frac{f(x+\beta\delta) - f(x)}{\delta^{\alpha+1}} d\delta\,.$$

Introducing a designation for the Marchaud fractional derivative

$$\mathbf{D}_{\pm}^{\alpha} f(x) := \frac{\alpha}{\Gamma(1-\alpha)} \int_0^{\infty} d\xi \frac{f(x) - f(x \mp \xi)}{\xi^{\alpha+1}} >,\; 0 < \alpha < 1$$

and making transformations analogous to those ones in [15], we obtain a Fokker-Planck generalized equation

$$\begin{aligned} \frac{\partial p(x,t)}{\partial t} &= -\frac{\partial}{\partial x}[a(x,t)p(x,t)] + \frac{1}{2}\frac{\partial^2}{\partial x^2}[b^2(x,t)p(x,t)] - \\ &- a_L \beta^{\alpha} \frac{\Gamma(1-\alpha)}{\alpha} \mathbf{D}_{+}^{\alpha} p(x,t) - b_L \beta^{\alpha} \frac{\Gamma(1-\alpha)}{\alpha} \mathbf{D}_{-}^{\alpha} p(x,t), \end{aligned} \tag{8}$$

which contains terms with fractional derivatives together with conventional terms. In case of good functions fractional derivatives on a line and Marchaud derivatives coincide $\mathbf{D}_{\pm}^{\alpha} f(x) = D_{\pm}^{\alpha} f(x)$ and according to equation (8) at $a_L = b_L$ terms with a fractional derivative coincide with analogous terms in work [11].

The represented derivation has shown strength and universality of the approach in terms of non-Wiener stochastic differential equations and extended an ordinary derivation of Fokker-Planck equation in case of the Gaussian process [1] to the Levy non-Gaussian processes with parameter $0 < \alpha < 1$. It is worth noting that the Levy measure can also be obtained from the rescaling procedure of noise term analogous to the one presented in [16,17].

Physical effects and regularities described by terms with fractional derivatives have not been discussed in this paper because they are analogous to the investigated ones in [11]. The main results of the paper consist in the demonstration of the unified derivation of kinetic equation for different cases from the non-Wiener SDE only, including the justification of terms with fractional derivatives.

Acknowledgments
This work is supported in part by the Russian Foundation for Basic Research (grant No. 13-02-00199-a).

5. References

[1] Gardiner C W 2004 *Handbook of stochastic methods for physics, chemistry and the natural sciences* (Berlin, Heidelberg: Springer-Verlag)
[2] Gardiner C W and Zoller P 2000 *Quantum noise* (Berlin, Heidelberg: Springer-Verlag)
[3] Maimistov A I and Basharov A M 1999 *Nonlinear optical waves* (Dordrecht: Kluwer Academic)
[4] Basharov A M 2011 *Phys. Lett.* A **375** 784
[5] Basharov A M 2011 *Phys. Lett.* A **375** 2249
[6] Basharov A M 2011 *Phys. Rev.* A **84** 013801
[7] Basharov A M 2011 *Phys. Lett.* A **376** 1881
[8] Metzler R and Klafter J 2000 *Phys. Rep.* **339** 1
[9] Zaslavsky G M 2002 *Phys. Rep.* **371** 469
[10] Barkai E, Naumov A V, Vainer Yu G, Bauer M and Kador L 2004 *J. Lumin.* **107** 21
[11] Chechkin A V and Gonchar V Yu 2000 *Zh.Eksp.Teor.Fiz.* **118** 730
[12] Chandrasekhar S 1943 *Rev. Mod. Phys.* **15** 1
[13] Kyprianou A 2006 *Introductory lectures on fluctuations of Levy processes with applications* (Berlin, Heidelberg: Springer-Verlag)
[14] Jacobs K 2010 *Stochastic Processes for Physicists* (Cambridge: CUP)
[15] Basharov A M 2012 *Nanosystems: physics, chemistry, mathematics* **3** 47
[16] Eliazar I I and Shlesinger M F 2012 *J. Phys.* A **45** 162002
[17] Eliazar I I and Shlesinger M F 2013 *Phys. Rep.* **527** 101

Breather-like pulses in a reduced Maxwell-Duffing model

E V Kazantseva[1,2], **A I Maimistov**[2,3],

[1] Joint Institute for High Temperatures, Russian Academy of Sciences, Izhorskaya 13/2, Moscow, Russia, 125412
[2] National Research Nuclear University "MEPhI", Kashirskoye sh., 31, Moscow, Russia, 115409
[3] Moscow Institute of Physics and Technology (State University), Institutskii per., 9, Dolgoprudnyi, Moscow region, Russia, 141700

E-mail: elena.kazantseva@gmail.com

Abstract. Propagation of pulses with oscillating amplitude of the electromagnetic field (breathers) is considered in the framework of a model in which the medium is represented by anharmonic oscillators with cubic nonlinearities. Numerical simulations demonstrate that the high frequency breathers propagate as the steady-state solutions and are very robust against collisions with one-soliton solutions of this model. Evolution of low-frequency breather leads to dispersion spreading and conversion the pulse into quasiharmonic wave.

1. Introduction

Nowadays the methods of generation of the pulses consisting of one or just a few cycles of the electromagnetic field are well established [1-7]. These pulses are called the extremely short pulses (ESPs). In the papers [8-9] the analytic solutions for a system of Maxwell-Bloch equations were found, these solutions are obtained without an assumption about slowness of variation of the envelope of the electromagnetic pulse. In general, the wave equation possesses solutions, corresponding to oppositely propagating waves. However, if the polarization response is small, and backward scattering could be neglected, the unidirectional approximation might be used instead. This approximation allows to reduce the wave equation to the first-order one without setting the limitations on the duration of the pulse. The nonlinear dynamics of the medium under electromagnetic pulse action is often modeled by an ensemble of anharmonic oscillators. For example, the propagation of linearly polarized ESP in the cubic nonlinear medium (described by Duffing model) is considered in [2,10]. Evolution of ESP in dispersive nonlinear medium modeled with quadratic nonlinear oscillators is considered in [1], the case without dispersion is considered in [11]. In these works the oscillator characterizes high-frequency (electronic) response of the medium on action of the electromagnetic pulse. As the ESP spectrum lies in the low-frequency range, the low-frequency oscillations of the ion lattice could contribute significantly to medium polarization. Propagation of the femtosecond pulse in a medium with nonlinearity caused both by electronic and ionic vibrations is studied in [12]. In this paper the propagation of linearly polarized breathers in a nonlinear dispersive medium represented by anharmonic oscillators with cubic nonlinearities is considered. The oscillator represents a response of the high-frequency electron degree of freedom to the electromagnetic field. A model of the propagation of breathers in a medium with cubic nonlinearity is analyzed by numerical computation. A breather solution of modified Korteweg-de Vries equation has been taken as the initial condition. We found that the low-frequency breathers are quickly destroyed by dispersion spreading and results in periodic quasiharmonic wave. Variation of parameters increasing breather's frequency leads to pulse stabilization. It is shown that the high-frequency breather evolution is very similar to behavior of the breather solutions in the completely integrable models. Moreover, an example of strong interaction illustrated by collisions of one-soliton solution of the Duffing model with the high-frequency breather, shows the stability of such a breather pulse.

2. System of Maxwell-Duffing equations

Evolution of the electric component E of the electromagnetic pulse in cubic nonlinear medium with polarization P of a unit volume of the medium is described by a wave equation

$$\frac{\partial^2 E}{\partial z^2}+\frac{1}{c^2}\frac{\partial^2 E}{\partial t^2}=\frac{4\pi}{c^2}\frac{\partial^2 P}{\partial^2 t}. \quad (1)$$

In the unidirectional approximation [13-14] it reduces to first-order equation

$$\frac{\partial E}{\partial z}+\frac{1}{c}\frac{\partial E}{\partial t}=-\frac{2\pi}{c}\frac{\partial P}{\partial t}.$$

To describe polarization response of nonlinear medium the oscillator model is used [15-17]:

$$\frac{\partial^2 X}{\partial t^2}+\omega_0^2 X+k_3 x^3=\frac{e}{m_{eff}}E(z,t). \quad (2)$$

Then polarization of a unit volume of the medium is defined as $P = n_A eX$, X - oscillator coordinate. Here n_A - density of atoms, e - electron charge, $m_{eff} = 3m/(\varepsilon + 2)$ - effective mass of an electron, ω_0 - oscillator eigenfrequency, k_3 - nonlinearity coefficient.

Let us introduce the variables $t = z/l$, $x = \omega_0(t - z/l)$ and fields $e = E/A_0$, $q = X/X_0$. The normalizing coefficients are $A_0 = m_{eff}\omega_0^2 X_0/e = m_{eff}\omega_0^3 e^{-1}(2\mu/|k_3|)^{1/2}$, $X_0 = (2\mu\omega_0^2/|k_3|)^{1/2}$, $l^{-1} = 2\pi n_A e^2/(m_{eff}c\omega_0) = \omega_p^2/2c\omega_0$, plasma frequency $\omega_p = (4\pi n_A e^2/m_{eff})^{-1/2}$, $\mu = k_3 X_0^2/2\omega_0^2$. The normalized system of Maxwell-Duffing equations in the unidirectional approximation, i.e. reduced Maxwell-Duffing (RMD) model becomes following:

$$\frac{\partial e}{\partial t}=-\frac{\partial q}{\partial x},\frac{\partial^2 q}{\partial x^2}+q+2\mu q^3 = e.$$

3. Numerical simulation

In the numerical simulation the following system of equation was studied:

$$\frac{\partial e}{\partial t}=-p,\ \frac{\partial q}{\partial x}=p,\ \frac{\partial p}{\partial x}=e-q-2\mu q^3.$$

The subject of the present study was numerical simulation of the breather-like pulses formation in the RMD model. To find the breather-like solitary wave in a RMD model the breather solution of the modified Korteweg-de Vries (mKdV) equation was employed as the initial condition:

$$e(x,t=0)=-\frac{4\beta}{\alpha_1}\frac{\alpha_1\cos\theta_2 ch\theta_1-\beta sh\theta_1\sin\theta_2}{ch^2\theta_1+(\frac{\beta}{\alpha_1})^2\sin^2\theta_2}, \quad (3)$$

here

$$\theta_1 = 2\beta(x-x_{10})+8\beta(\beta^2-3\alpha_1^2-0.25)t, \quad (4)$$

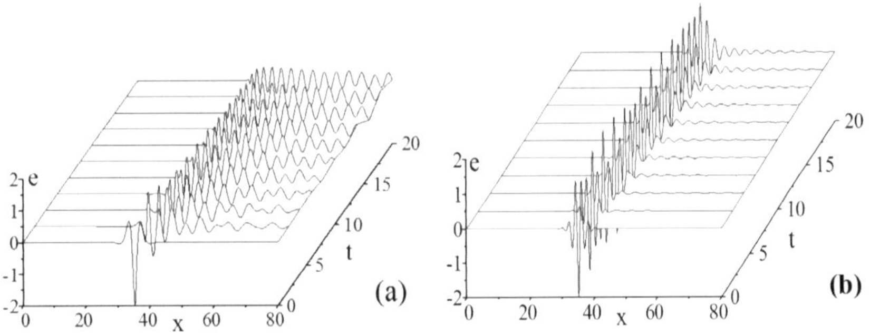

Figure 1. Evolution of a mKdV breather in a RMD model. Figure (a) shows evolution of a low-frequency breather characterized by the parameter $\alpha_1 = 0.5$, figure (b) corresponds to $\alpha_1 = 1.5$.

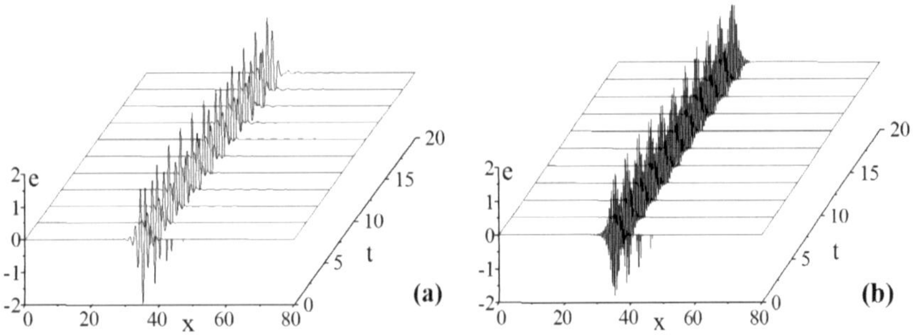

Figure 2. Evolution of a mKdV breather in a RMD model. Figure (a) shows evolution of a low-frequency breather characterized by the parameter $\alpha_1 = 2$, figure (b) corresponds to $\alpha_1 = 5$.

$$\theta_2 = 2\alpha(x - x_{20}) + 8\alpha_1(\alpha^2 - 3\beta^2 + 0.25)t. \tag{5}$$

The parameter $\beta = 0.5$, α_1 which defines breather's frequency is varied.

The first figure illustrates spreading of the low-frequency mKdV breathers in a RMD model. Dispersion leads to broadening of the initial pulse and transformation into quasiperiodic waves. The localized breather-like pulses do not form here. Increase for the parameter α_1, which increases the frequency of the initial pulse, leads to stabilization of the wave packet, it is illustrated by the figure 2. Collision of such high-frequency pulses with steady-state solutions of the RMD model demonstrates their stability (figure 3).

An initial high-frequency pulse obtained by modulation of the steady-state pulse of the RMD model with harmonic wave shows considerable robustness during its evolution (figure 4), which is similar to that for high-frequency mKdV breathers evolving in RMD model. The frequency spectra of such pulses is located at the frequency of the carrier wave, it remains almost unchanged under pulse propagation. These modulated pulses also show stability against collisions with steady state pulses of RMD model.

4. Conclusion

The propagation of the breather-like pulses in a cubic nonlinear medium was studied numerically. The initial conditions were chosen corresponding to breather solutions of mKdV equation. It is shown that low-frequency breather-like pulses transform into periodic quasiharmonic waves during propagation due to dispersion effect. Increase for the frequency of the initial pulse leads

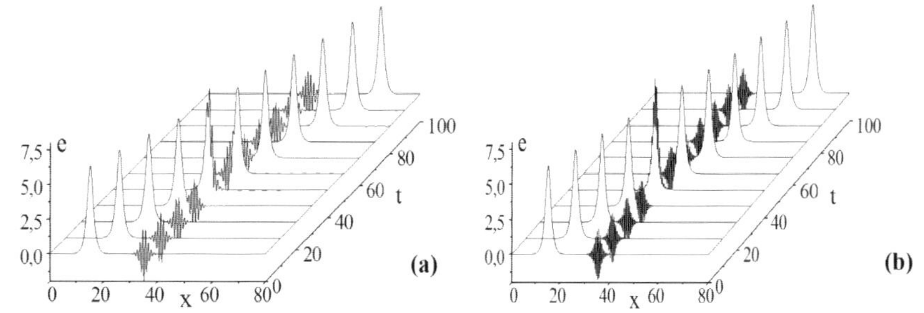

Figure 3. Collision of a RMD solitary wave (characterized by $\alpha = 3$, $\mu = 0.1$) with a high-frequency pulse evolved from mKdV breather. Figure (a) shows evolution of a low-frequency breather characterized by the parameter $\alpha_1 = 3$, figure (b) corresponds to $\alpha_1 = 5$.

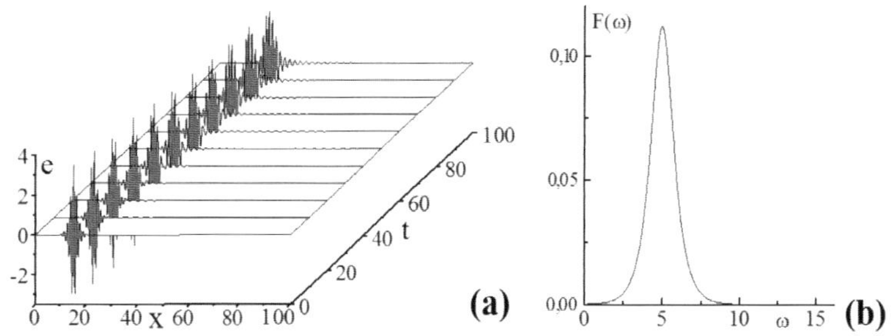

Figure 4. Evolution of a pulse set as $e(x,0) = \alpha\sqrt{\frac{\alpha-1}{\mu}}sech(\sqrt{(\alpha-1)}(x - \frac{t}{\alpha} - x_0))\cos 5(x - x_0)$ at $\alpha = 2$, $\mu = 0.3$. The Fourier spectra of the pulse doesn't change during propagation (figure (b)).

to its stabilization, these high-frequency waves are stable against collisions with steady state solitary waves of RMD model.

Acknowledgments
The authors thank the organizing committee of the scientific school for hospitality.

References

[1] Dubrovskaya O B and Sukhorukov A P *Izv. RAN, ser. Phys.* 1992 **56** 184
[2] Kaplan A E *Phys. Rev. Lett.* 1994 **73** 1243
[3] Kaplan A E and Shkolnikov P L *Phys. Rev. Lett.* 1995 **75** 2316
[4] Sazonov S V and Trifonov E V *J. Phys. B* 1994 **27** L7
[5] Sazonov S V and Yakupova L S *J. Phys. B* 1994 **27** 369
[6] Kaplan A E, Straub S F and Shkolnikov P L *Opt. Lett. Lett.* 1997 **22** 405
[7] Agrawal G P 1995 *Nonlinear Fiber Optics* (Academic Press: San Diego)
[8] Akimoto K *J. Phys. Soc. Japn.* 1996 **65** 2020
[9] Andreev A V *Phys. Lett. A* 1993 **179** 23
[10] Maimistov A I and Elyutin S O *J. Mod. Opt.* 1992 **39** 2201
[11] Akopyan A A and Oganesyan D L *Quant. Electron.* 1998 **25** 954 (in Russian)
[12] Kozlov S A and Sazonov S V *JETP* 1997 **84** 221
[13] Bullough R K, Jack P M,. Kitchenside P W, and Saunders R. *Phys. Scripta* 1979 **20** 364
[14] Caudrey P J, Eilbeck J C, Gibbon J D, and Bullough R K *J. Phys. A* 1973 **6** 1337
[15] Bloembergen N 1965 *Nonlinear Optics* (Benjamin: New York).
[16] Yariv A 1967 *Quantum Electronics* (John Wiley and Sons, Inc.: New York and London)
[17] Butcher P N, and Cotter D 1990 *The Elements of Nonlinear Optics* (Cambridge University Press)

Methods of characterization of synthetic opal films

A V Koryukin, A A Akhmadeev and M Kh Salakhov

Department of Physics, Kazan Federal University, 18 Kremlevskaya St, Kazan 420008, Russia

E-mail: Artem.Koryukin@kpfu.ru

Abstract. We developed methods for determination of thickness, number of layers and filling fraction of silica particles for synthetic opals. We show that the filling fraction is considerably less than for ideal close-packed structure, which is important for practical and theoretical applications.

1. Introduction

Photonic crystals (PC) are materials with a spatial periodicity in their dielectric function (or, equivalently, a periodical index of refraction) [1]. PC can have a photonic bandgap, i.e. a frequency window in which propagation of light through the crystal is difficult [2]. The inhibition of light propagation makes it possible to control spontaneous emission [3] that has many important applications [4]. Recently it has been shown [5] that the modification of the interaction of an electron with its own radiation field in the PC medium in fact is equivalent to that the electron mass changes its value and this opens up new vistas for their practical applications. This effect, for example, can in turn give rise to the significant modification of the strong interaction of emitters in PCs with a resonant laser field, predicted in [6]. At present there are numerous approaches to the creation of PC by using the lithography [7], interferential holography [8] and self-assembly of colloidal particles [9]. We use self-assembly of colloidal silica particles (SiO_2). It forms synthetic opal films consisting of close-packed spherical particles of silica. The photonic crystal samples based on (SiO_2) were prepared by vertical deposition onto a glass substrate by capillary action [9]. The process is quite simple in terms of hardware design and have no fundamental restrictions both on the sample size and the number of photonic crystals produced in one cycle of synthesis. So our purpose is to investigate methods for quick and easy determination of parameters such as thickness, number of layers and filling fraction of silica particles. Such methods and parameters required at multiple syntheses for different experiments with PC. Effective refractive index, thickness of layer and filling fraction of silica particles can be determined by spectrum position of band gap. Thickness of PC and number of layers can be determined by different methods depending on the number of layers.

2. Spectroscopic methods of parameters determination

Effective refractive index, thickness of layer and filling fraction of silica particles can be determined by transmission or reflection spectra measured at different incidence angles with the help of Bragg's law. Taking account of Snell's law, it is written as:

$$\lambda = 2d_{(111)}\left(n_{eff}^2 - \sin^2\theta\right)^{1/2}, \tag{1}$$

where λ is the wavelength corresponding to the bandgap, $d_{(111)}$ is the period of the structure (the distance between the (111) planes, so-called thickness of layer), n_{eff} is the effective refractive index, θ is the angle between the normal to the sample surface and the direction of the incident light.

Figure 1 shows the transmission and reflection spectra of PC samples based on silica. The dependence of band gap spectral position (λ) on the incidence light angle (θ) is linear in terms of $\lambda^2 - \sin^2\theta$ (figure 2).

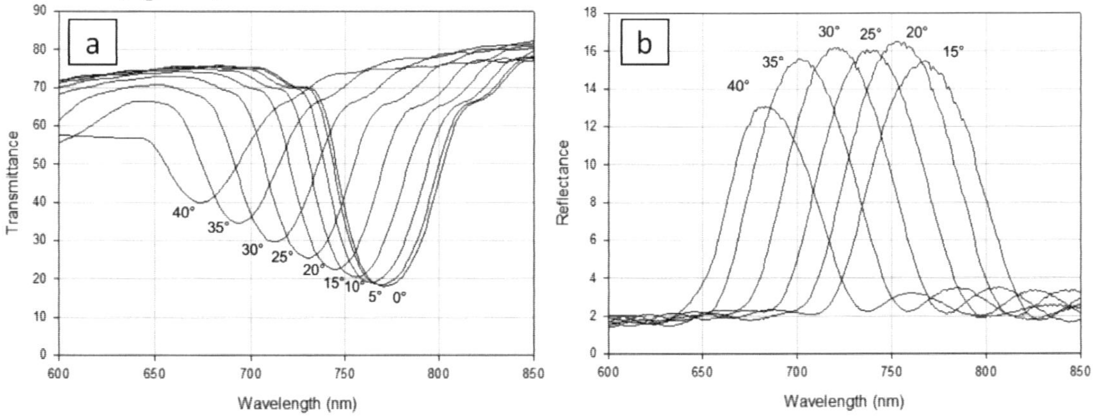

Figure 1. Transmission (a) and reflection (b) spectra of the PC samples.

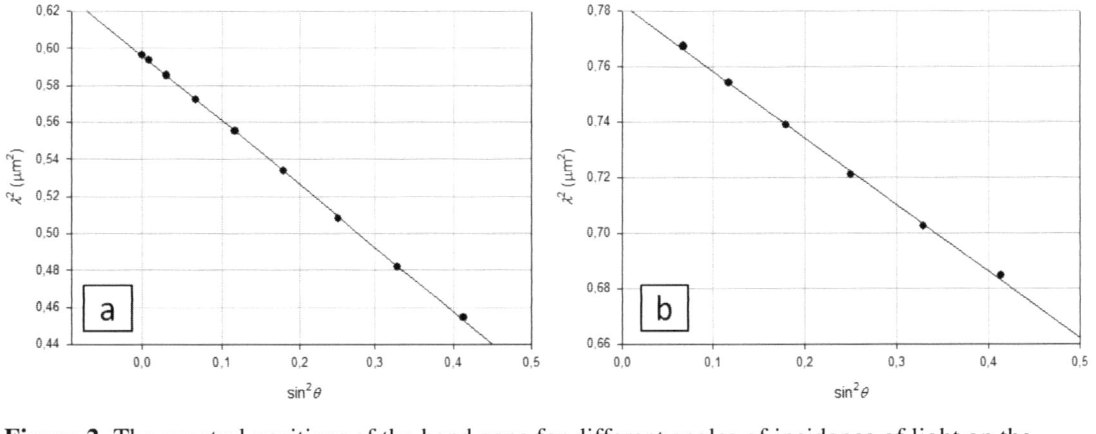

Figure 2. The spectral positions of the band gaps for different angles of incidence of light on the sample surface: a) for the transmission spectrum; b) for the reflection spectrum.

Using a linear least-squares approximation, we have calculated the effective refractive index as well as the period of the structure. The effective refractive index of the PC is

$$n_{eff} = \left(n^2 \cdot f + n_{air}^2 \cdot (1-f)\right)^{1/2}, \tag{2}$$

where n is the refractive index of the SiO_2 particles (n = 1.46), n_{air} is the refractive index of air (n_{air} = 1), f is the filling fraction of silica particles. According to close-packing of equal spheres the f parameter is equal to 0.7405 and n_{eff}, in turn, is 1.36. Using the equation 2, the filling fraction of silica particles can be calculated as follows:

$$f = \frac{n_{eff}^2 - n_{air}^2}{n^2 - n_{air}^2}. \tag{3}$$

The calculation results for several different samples showed that the filling fraction lies in the range from 0.59 to 0.66, while the average value of $d_{(111)}$ and n_{eff} was 280 nm and 1.30, respectively. Filling fraction obtained for the PC samples is significantly less than for the ideal structure with close-packing of identical spheres. Since the defects of lattice affect a packing density, conventional PC samples have a larger volume of cavities.

3. Thickness and number of layers for thick films

Thickness of the thick films (number of layers greater than about 20) can be determined by microscopy of cross-section or measurement the mass of the samples. Microscopy of PCs cross-section is the obvious way for thickness determination using the image from optical microscope which is shown on figure 3 (a).

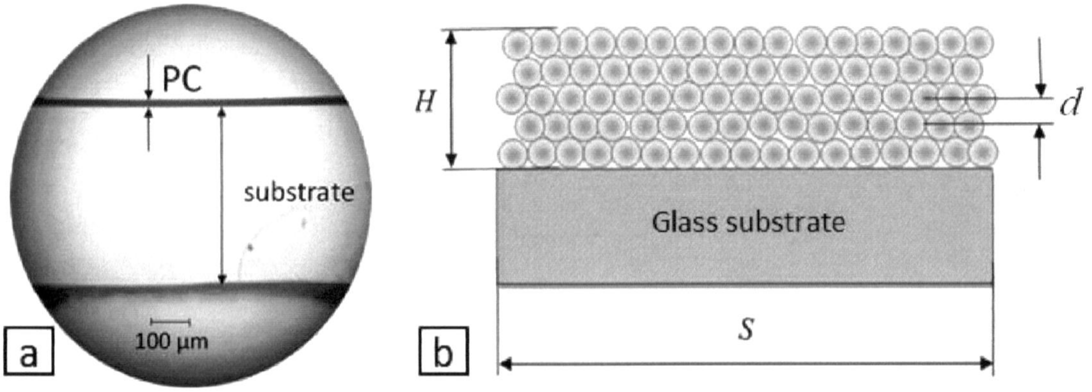

Figure 3. a) Snapshot of PCs cross-section by using optical microscope; b) scheme of PCs cross-section.

Thickness of PC can be determined via mass of samples:

$$H = \frac{m}{\rho \cdot S \cdot f}, \tag{4}$$

where H is the thickness of PC, m is the mass of the sample, ρ is the density of silica particles, S is the square of PC surface. Number of layers (N_L) can be determined via thickness of PC and thickness of the layer:

$$N_L = H / d_{(111)}. \tag{5}$$

The calculation results for one of the samples were as follows: m = 17 mg, N_L = 26 and H = 17.3 μm. The proposed method for determining the thickness cannot be fine applied to a relatively thin film, because their mass is so small that it becomes comparable with the measurement error.

4. Thickness and number of layers for thin films

For the thin PC films number of layers can be determined by oscillations on the transmission spectrum (figure 4). The interference of light reflected by the top and bottom surfaces causes the well-known Fabry-Perot interferences [10].

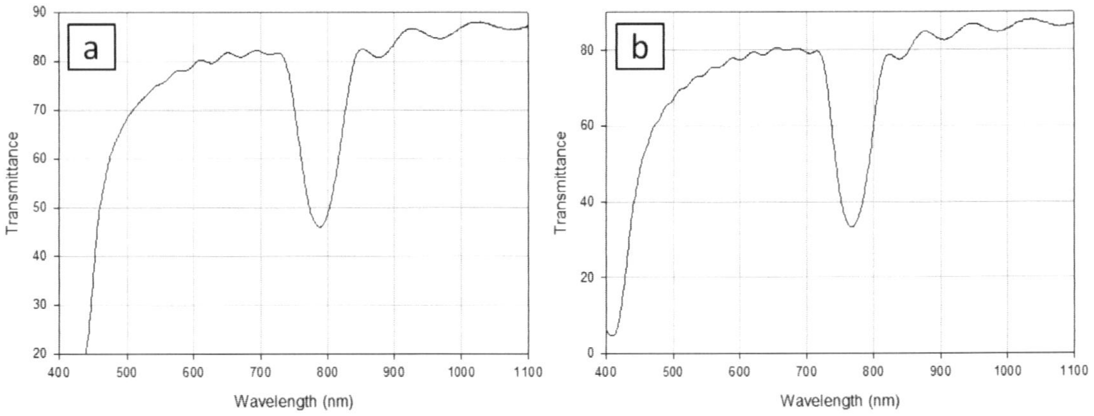

Figure 4. Transmission spectra of thin PC films: a) 12 layers; b) 15 layers.

Combining the Bragg's law and interference conditions we obtain the following expression:

$$N_L = \frac{p \cdot \lambda_0 \cdot \lambda_p}{\lambda_{Br}\left(\lambda_0 - \lambda_p\right)}, \tag{6}$$

where λ_0, λ_p is wavelengths corresponding to different fringes, p is a positive integer numbering consecutive minima between the fringes, λ_{Br} – the Bragg diffraction wavelength (bandgap). Thickness of PC can be obtained via number of layers and thickness of layer: $H = N_L\, d_{(111)}$, where thickness of the layer $d_{(111)}$ is calculated by the method described above.

Thus, the thickness of the PC film for the sample in figure 4 (a) amounted to 3.6 μm, number of layers – 12. For the sample in figure 4 (b), these parameters were 4.4 μm and 15, respectively.

In the case of a film of thickness more than 8 μm, intensity oscillations cannot be clearly observed in the visible light. Thus, determination of number of layers for thick films by Fabry-Perot oscillations is not possible.

5. Conclusion

In summary, we have developed methods for quick and easy determination of synthetic opal's parameters such as thickness and number of layers for thick-film PCs. We have shown that filling fraction of silica particles obtained for conventional PC are considerably smaller than for ideal PC structure due to the presence of packing defects.

References

[1] Yablonovitch E 1987 *Phys. Rev. Lett.* **58** 2059–62
[2] Novotny L and Hecht B 2006 *Principles of nano-optics* (New York: Cambridge Univrsity Press) p 539
[3] Yamasaki T and Tsutsui T 1998 *Appl. Phys. Lett.* **72** 1957–59
[4] Yablonovitch E 1993 *J. Opt. Soc. Am. B* **10** 283–295
[5] Gainutdinov R K, Khamadeev M A and Salakhov M K 2012 *Phys. Rev. A* **85** 053836
[6] Gainutdinov R Kh and Mutygullina A A 2009 *Physica Scripta T* **135** 014012
[7] Yao P, Schneider G J, Prather D W, Wetzel E D and O'Brien D J 2005 *Optics Express* **13**(7) 2370–76
[8] Duneau M, Delyon F and Audier M 2004 *J. Appl. Phys.* **96**(5) 2428–36
[9] Vlasov Y A, Bo X-Z, Sturm J C and Norris D J 2001 *Nature* **414** 289–293
[10] Wang Jing, Yuan Chun-Wei and Tang Fang-Qiong 2005 *Chinese Phys.* **14** 1581

Journal of Physics: Conference Series **478** (2013) 012014 doi:10.1088/1742-6596/478/1/012014

Synthesis of nickel microwires with nanostructured surface for electrodes of chemical current sources

M V Morozov[1], A Kh Gilmutdinov[1] and M Kh Salakhov[2]

[1] Kazan State Technical University named after A.N. Tupolev, 10 K.Marx St., 420111 Kazan, Russian Federation
[2] Ibstitute of Dhysics, Kazan Federal University, 18 Kremlyovskaya St., 420008 Kazan, Russian Federation

e-mail: Morozov.ksu@inbox.ru

Abstract. Electrodes with a nanostructured surface based on nickel micro- and nanowires have wide application in magnetic materials and are tested in rechargeable batteries. For the synthesis of nickel wires a chemical method is used. A Ni electrode based on a Ni microwires with roughened surface was prepared and its electrochemical properties were investigated in electrode reaction for NiOOH.

1. Introduction

An increase in the charge/discharge rate of rechargeable batteries of more than one order of magnitude is required to meet the future demands of hybrid electric vehicles and efficient energy storage [1]. Limits exist because of the intrinsically slow diffusivity of the reactive ion in the solid state, which inevitably limits the rate of charge/discharge [1]. With carbon or carbon-buffered alloys as the anode a protecting and passivating solid-electrolyte interface (SEI) layer forms on the anode, this limits the safe charging rate of lithium-ion batteries [2]. There is also increasing interest in the possibility to use metal-based nanostructures instead of microstructures, with the goal of reducing the electrode volume, securing better safety, as well as achieving increased capacity [2].

NiOOH as cathode material for Ni-based batteries has a high theoretical capacity – 286 mAh/g, if the following electrode reaction takes place:

$$NiOOH + H_2O + e^- = Ni(OH)_2 + OH^- \qquad (1)$$

A serious drawback of NiOOH is its low electrical conductivity [3]. The way of optimization of NiOOH electrode has directed to reach high rate capability of charge/discharge processes.

The highest capacity and good cycling behavior for NiOOH cathodes were observed for nickel sintered and nickel foam electrodes. The highest current density typically used for Ni sintered current collectors are about 2-3 mA/cm^2. Because NiOOH has a low electrical conductivity a high rate capability is only provided by using conductive additives like nickel [4]. But to the best of our knowledge nickel micro- and nanowires with roughened surface were not investigated yet as base material.

In technical applications nickel sintered plaque is used as current collector, because of its high specific surface area and the many routes towards the design of structured electrode. Nickel shows lower resistance compared to carbon – 69.3 nΩ^{-1}m^{-1} vs. 7,800 nΩ^{-1}m^{-1} for graphite. Nickel as current collector for positive electrodes of rechargeable batteries has good electrical conductivity in terms of weight and price [5]. So, nickel micro- and nanowires are well suited scaffolds for the NiOOH electrode. In addition, nickel wires are also interesting from the technological point of view because the synthesis of nickel wires via chemical deposition offers low cost and large-scale production [6-13].

Here we describe the preparation of a Ni electrode for nickel batteries based on nickel micro- and nanowires. The aim of the investigation was to demonstrate the high current density for NiOOH cathodes by using nickel scaffolds. We focus on the preparation of nano-architectured electrodes based on nickel wires that are not expensive and can be applied in batteries.

2. Experimental

Ni wires were prepared by chemical deposition using the technique described in [12, 13] with the surfactant cetyl trimethylammonium bromide (CTAB) and polyethylene glycol (PEG, M=10000). 0.660 g $NiCl_2*6H_2O$ (Acros organics, 2,8 mmol), 0.8 g PEG (Alfa Aesar, 0.08 mmol), 0.3 g CTAB (Acros organics, 1.1 mmol) and 45 ml distilled water were mixed into a 100 ml round-bottom flask at room temperature under stirring with magnetic stirrer. A grass-green color was observed in the solution. Then 5 ml of 80 wt % hydrazine hydrate solution was added dropwise into the solution under stirring with magnetic stirrer, and the color changed from grass-green to navy blue. The round-bottom flask with reaction mixture was closed tightly by tap and was placed into water thermostat at temperature 70 $^{\circ}$C for 3 h. After the reaction was completed, the resulting black solid product floated to the top of the flask, indicating the formation of metallic nickel. The product was washed with distilled water and absolute isopropyl alcohol several times. For the control of synthesis magnetic field with magnitude between 0.05 and 0.86 T was used.

The structure and morphology of these products were characterized by X-ray diffraction (XRD), atomic force microscopy (AFM) and scanning electron microscopy (SEM). XRD measurements were carried out on an XRD-7000 diffractometer from SHIMADZU. SEM images were taken with a Auriga Crossbeam (FIB-SEM) workstation microscope from Zeiss. The specific surface area of the wires was determined by Brunauer-Emmet-Teller (BET) method on an Autosorb-iQ MP surface area analyzer from Quantachrome. For AFM investigation Bruker Innova atomic force microscope was used.

The Ni/NiOOH electrodes with and without wires were tested in three electrode electrochemical cells. The prepared electrodes were used as working electrodes and Pt electrode was used as counter electrode. The reference electrode was Ag/AgCl. Prepared electrodes were tested in 0.1 M NaOH electrolyte solution. The specific feature of Ni/NiOOH wire electrode is dendrite structure with roughened surface compared to the sintered and foam electrode.

Cyclic voltammetry measurements of the electrodes were performed on a PGSTAT 302 N potentiostat/galvanostat from Metrohm Autolab.

3. Results of synthesis and characterization of nano-architectured electrode based on a nickel wires with roughened surface

In figure 1 microstructures of the Ni/NiOOH sintered and wires electrodes surface are shown. In figure 2 Ni individual wires are shown. Influence by magnetic field during synthesizing of the wires causes a roughened surface with nanocones. The nanocones are about 200 nm – 750 nm in height, and wires do not stick together. The morphology of the prepared microwires agrees well with experimental data related to the structure of the Ni wires surface after hydrothermal synthesis [7, 8]. It can be seen from AFM images that the Ni nanocones distribute on the Ni electrodes surface.

Journal of Physics: Conference Series **478** (2013) 012014 doi:10.1088/1742-6596/478/1/012014

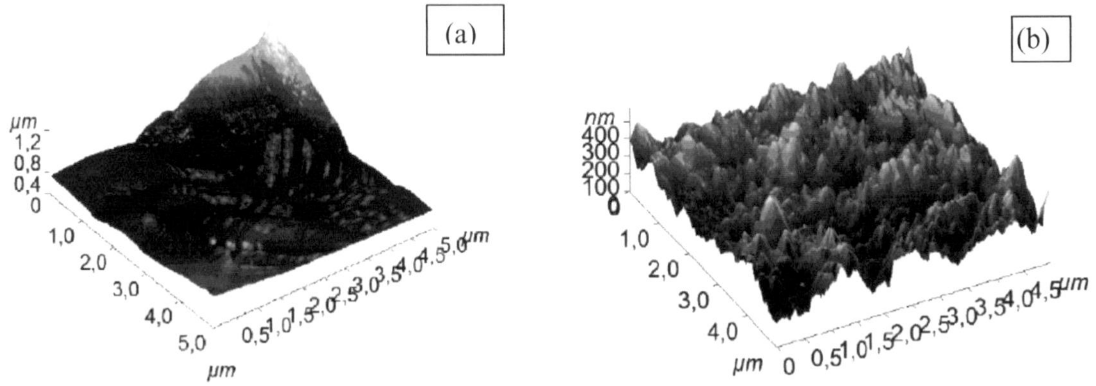

Figure 1. AFM images of microstructure of the surface of: a) Ni sintered electrode; b) Ni wires electrode with roughened surface.

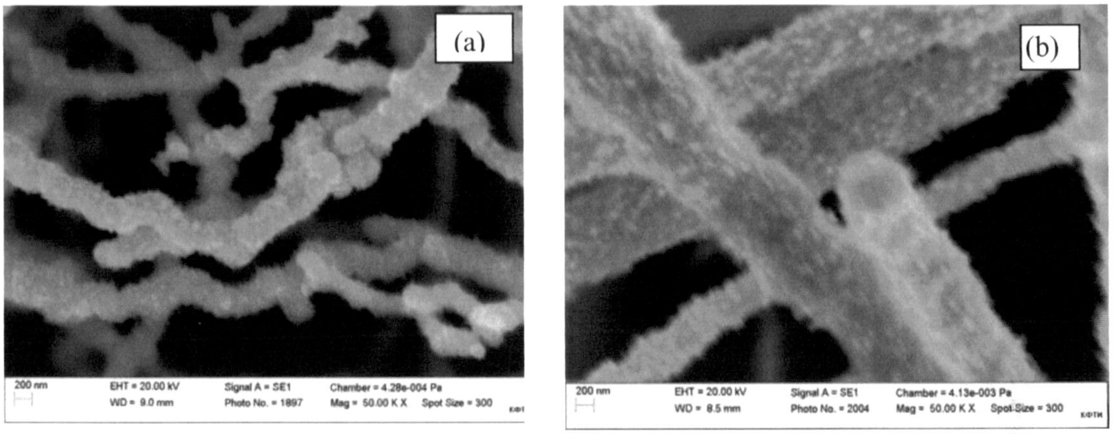

Figure 2. SEM images of individual nickel wires synthesized: a) without the magnetic field; b) under the influence of magnetic field 0.06 T.

AFM data were used for the analysis of roughness of the nickel electrodes (figure 3). The advantage of AFM investigation is 3D characterization of the electrodes surface. For the numerical analysis of roughness fractal dimension was used. From the analysis of AFM images it can be concluded that with increasing of roughness of the surface fractal dimension has increased. Fractal dimension for the sintered electrode is 2.27, fractal dimension for the nickel wires electrode with roughened surface is 2.55.

For the determination of chemical purity and crystal structure of the electrodes XRD measurements were carried out. On the spectra (figure 4) three main peaks were found and indexed by crystallographic planes of Ni: 57.6° - (111), 67.4° - (200), 102.2° - (220). There are no any impurities like oxides and hydroxides, that means high purity and crystal structure of nickel electrodes. Size of crystallites from Sherer equation was found about 40 nm, which corresponds to the data in [13].

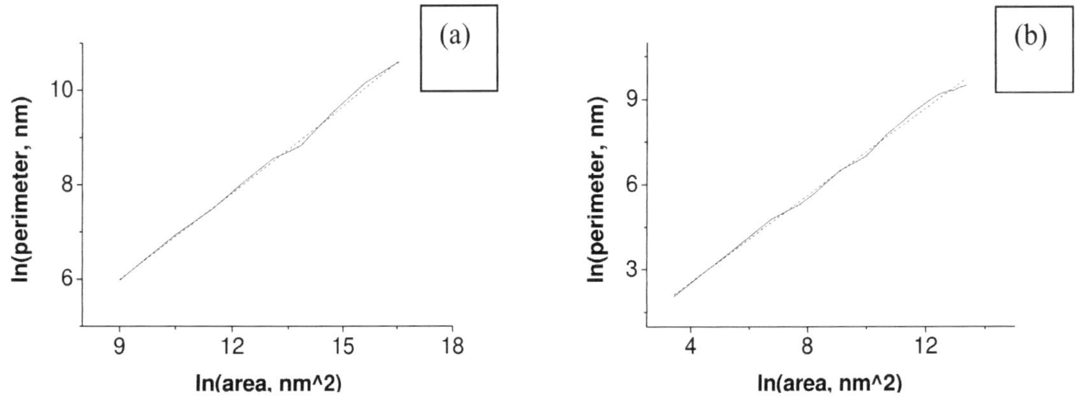

Figure 3. Dependence of perimeter on the area of the islands on the nickel electrodes surface in logarithmic coordinates: a) nickel sintered electrode; b) Ni wires electrode with roughened surface.

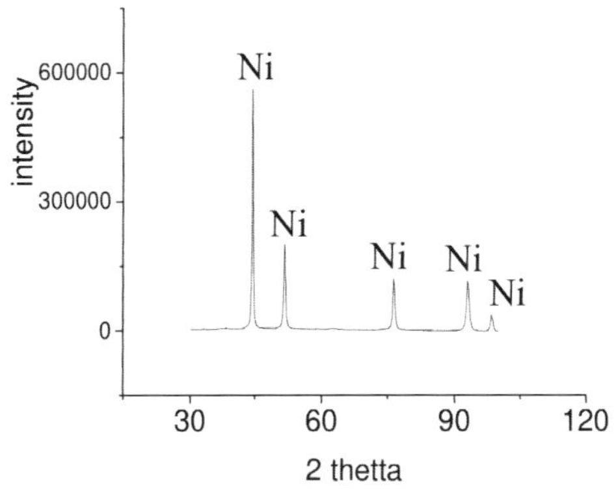

Figure 4. XRD measurements of synthesized nickel wires (λ=0.193728 nm).

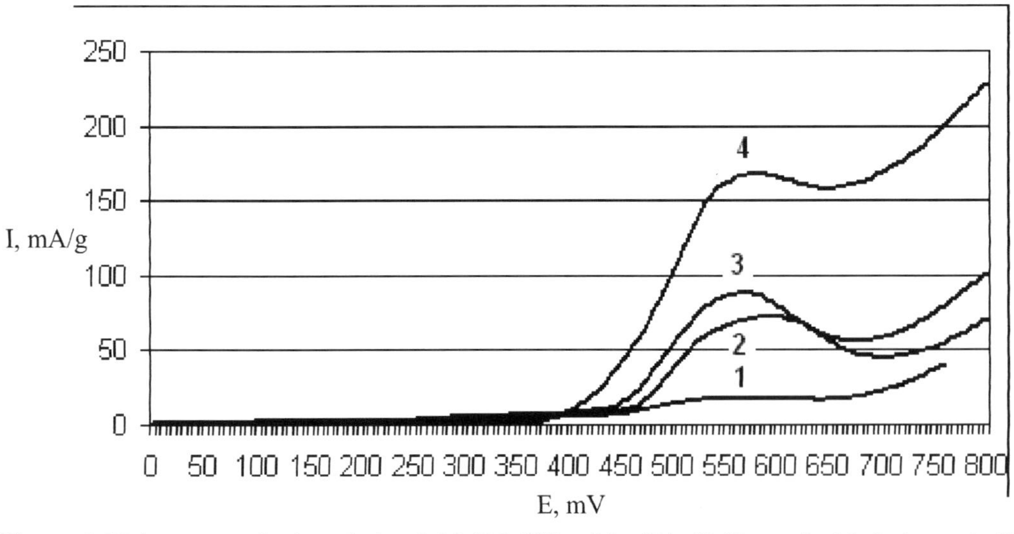

Figure 5. Voltammetry in the solution 0.1 M NaOH at 20 mV/s. 1) Sintered nickel electrode (SNI); 2) SNI with microwires without nanocones; 3) SNI with microwires with nanocones; 4) SNI with submicrowires (D=330 nm) with nanocones.

For the investigation of Ni/NiOOH electrode current density we analyzed cyclic voltammetry data for the wires electrode (figure 5). We observed a linear dependence of peak current vs. the square root of the voltage scan rate. The electrode shows stable performance at scan rates from 5 to 60 mV/s. The specific surface area of nickel submicrowires with nanocones from BET measurements is 5.5 m^2/g. It is more than one order higher than for nickel sintered electrode, that express in high current density of nanostructured electrodes. The high current density capabilities of the Ni/NiOOH electrode are related to roughened nickel wires surface that acts as effective current collector.

Investigation of the influence of magnetic field on current density of the electrodes demonstrates the increasing of current density with increasing of magnetic field (figure 6). It was found that increasing of magnetic field result in increase of the electrodes surface area [14]. This proves that we see indeed the effect of the surface area on the current density of the electrodes. To compare the current density of synthesized nanostrucutred electrode with Ni/NiOOH sintered electrode we presented experimental data in the table 1.

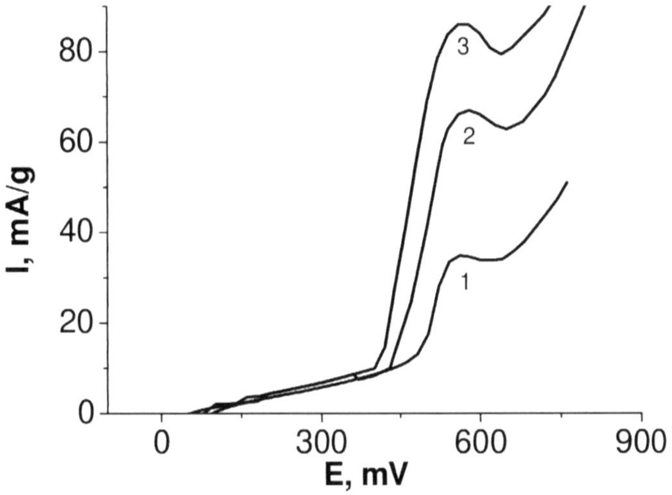

Figure 6. Voltammetry in the solution 0.1 M NaOH at 20 mV/s. 1) nickel foam electrode; 2) nickel foam electrode modified nanocones; 3) nickel foam electrode modified nanocones under the influence of magnetic field.

Table 1. Comparison of current density for foam Ni/NiOOH electrode and foam Ni/NiOOH electrode with roughened surface.

Parameter	Foam Ni/NiOOH electrode	Foam Ni/NiOOH electrode with roughened surface
Current density per mass, mA/g	34±3	66±7
Capacity density per surface area, mA/cm^2	3.3±0.3	6.3±0.5

4. Conclusion

The new approach of using nickel nanostructures as base material for NiOOH electrodes as cathode for Ni-based batteries is presented. The electrodes with Ni microwires with roughened surface were prepared by chemical deposition. The electrochemical properties of the Ni/NiOOH wires electrode are investigated and the results reveal a high current density performance. This is explained by effective charge transport because of using nickel wires core as current collector. So, we presented an interesting way of preparing NiOOH electrodes that give fundamental understanding of the mechanism how NiOOH cathodes work. The methods of the synthesis of Ni wires and the preparation of the electrode presented here are cheap, so it can be useful for preparation of NiOOH electrodes with high energy density.

Acknowledgments
This work is supported by RFBR, project № 12-02-97050-r_Volga region_a.

References
[1] Bruce P J, Scrosati B and Tarascon J M 2008 Nanomaterials for rechargeable lithium batteries *Angew. Chem. Int. Ed.* **47** 2930-46
[2] Nazri G and Pistoia G 2003 *Lithium Batteries: Science and Technology* (Berlin: Springer)
[3] Shukla A K, Venugopalan S and Hariprakash B 2001 Nickel-based rechargeable batteries *Journal of Power Sources* **100** 125-48
[4] Yang Q M, Ettel V A, Babjak J, Charles D K and Mosoiu M A 2003 Pasted $Ni(OH)_2$ electrodes using Ni powders for high-drain-rate, Ni-based batteries *J. of Electrochem. Society* **150** A543-50
[5] Whitehead A H and Schreiber M 2005 Current collectors for positive electrodes of lithium-based batteries *J. of Electrochem. Society* **152** A2105-13
[6] Liu C, Guo L, Wang R, Deng Y, Xu H and Yang S 2004 Magnetic nanochains of metal formed by assembly of small nanoparticles *Chemical Communications* **23** 2726-27
[7] Chou K S, Ren C Y and Lo C T 2002 Method of synthesizing nickel fibers and the nickel fibers so prepared *US Patent* **B1** 6.375.703
[8] Shui X and Chung D L 2000 Submicron diameter nickel filaments and their polymer-matrix composites *Journal of Materials Science* **35** 1773-85
[9] Morozov M V, Batalin G A, Gareev B I, Gilmutdinov A Kh and Salakhov M Kh 2011 Synthesis of nickel fibers with the roughened surface without influence of magnetic field and surfactants and investigation of electrochemical activity of them *Nanotechnics* **4** 61-66
[10] Zhu W H, Flanzer M E and Tatarchuk B J 2002 Nickel-zinc accordion-fold batteries with microfibrous electrodes using a papermaking process *Journal of Power Sources* **112** 353-66
[11] Liu Y, Wang H, Li J, Lu Y, Wu H, Xue Q and Chen L 2007 Monolithic microfibrous nickel catalyst co-modified with ceria and alumina for miniature hydrogen production via ammonia decomposition *Applied Catalysis* A: *General* **328** 77–82
[12] Morozov M V, Batalin G A, Gareev B I and Gilmutdinov A Kh 2010 Method of synthesis nickel micro- and nanofibrous structures with high surface area *Nanotechnics* **3** 29-33
[13] Helin N, Qianwang C, Min N, Yaoshun J and Xiaojuan W 2004 Synthesis and one-dimensional self-assembly of acicular nickel nanocrystallites under magnetic fields *The Journal of Physical Chemistry* B **108** 3996-99
[14] Morozov M V, Batalin G A, Gareev B I, Gilmutdinov A Kh, Tagirov M S and Salakhov M Kh 2012 Investigation of the surface rougheness for the nickel fibers synthesized under an applied magnetic field *Nanotechnics* **2** 48-54

Optical near-field scattering in a tip-sample system: A Green's function approach

A M Rogov, A R Gazizov, S S Kharintsev and M Kh Salakhov

Department of Optics and Nanophotonics, Institute of Physics, Kazan Federal University, Kremlevskaya 16, 420008 Kazan, Russia

1arogov@kpfu.ru

Abstract. The optical near-field contribution to the total scattered intensity for a spherical metallic nanoparticle close to a dielectric planar interface is analytically calculated using a Green's function formalism. This system is suited for probing a specimen of interest on a dielectric substarte with a taper optical metallic nanoantenna. Based on the effective polarizability model we find a Green's function in the dipolar approximation, that allows one take into account the presence of the dielectric substrate. In particular, we provide evidence that the latter noticably influences the near-field contribution and it should necessarily be included in the generalized formalism for calculating the Green's functions for Raman scattering and fluorescence. A comparative analysis of the analytical model and finite-difference time domain based numerical simulation is performed.

1. Introduction

In classical optics an incident radiation polarizes an object under study and the induced polarization gives rise to emitted radiation through, for example, Rayleigh or Raman scattering. According to a near-field paradigm the interaction between optical radiation and matter is achieved by means of an intermediate object, often referred to as an *optical antenna, tip or probe*, and that is brought to the close proximity of the matter [1]. Light-matter interactions at the nanoscale have led to fast development of optical near-field microscopy methods, which allow one to probe a specimen with a sub-wavelength spatial resolution less than the Abbe diffraction limit and single molecule sensitivity due to the giant field enhancement [2,3]. Thus, optical antennas play a crucial role in near field optical spectroscopy and microscopy [4]. They represent nano-sized devices enabling to transfer freely propagating electromagnetic waves into localized modes and vice versa [5]. In general the near-field contributes to the total scattered field E_{sc}, but it rapidly vanishes in the far-field zone. Within the frameworks of the Green's function formalism, a tip-sample system can be represented as follows:

$$G = G_0 + G_{sc}, \tag{1}$$

where G_0 and G_{sc} are the Green's functions for freely propagating and scattered electromagnetic waves, respectively. The scattered field in the tip-sample system can be, in general, expanded to a series, namely:

$$E_{sc} = E_s + E_t + E_{st} + E_{ts} + E_{tst} + \dots \tag{2}$$

where E_s and E_t are fields emitted or scattered from a sample and a tip, respectively; E_{st} is the field scattered by the sample and then re-emitted by the tip; E_{st} is the scattered field from the sample under illumination coming from the tip [1]; E_{tst} is a contribution to the scattered field as shown in figure 1(a). As a rule the sample represents an object under study on a substrate (glass, mica, etc.). Novotny's group has analytically calculated a Green's function for tip-enhanced Raman scattering from a single walled carbon nanotube [6,7] (figure 1(b)). However, this approach does not take into account the Rayleigh scattering from the substrate. This is important for probing a specimen with a scattering-type antennas which allow one to localize and enhance incident optical radiation due to the lighting rod effect. Figure 1(c) demonstrates a general problem statement - a pointed metallic tip is brought to the dielectric substrate and is aside illuminated. Our aim is to determine a Green's function for near-field scattering with the tip-sample system and to compare it with the numerical simulation based on the finite-difference time domain (FDTD) method.

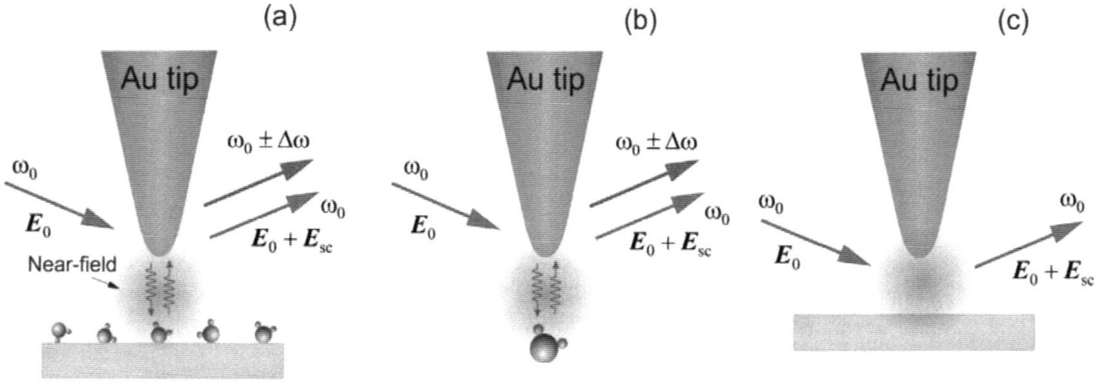

Figure 1. Schematic of optical radiation scattering in the tip-sample system.

2. Results and discussion

For finding the Green's function for Rayleigh scattering we replace the tapered nanoantenna (see figure 1) with a metallic nanosphere with a radius ρ and a permittivity ε_m, that is placed at the distance Δ from a planar interface between two dielectric media with permittivities ε_1 and ε_2, respectively. This is schematically shown in figure 2.

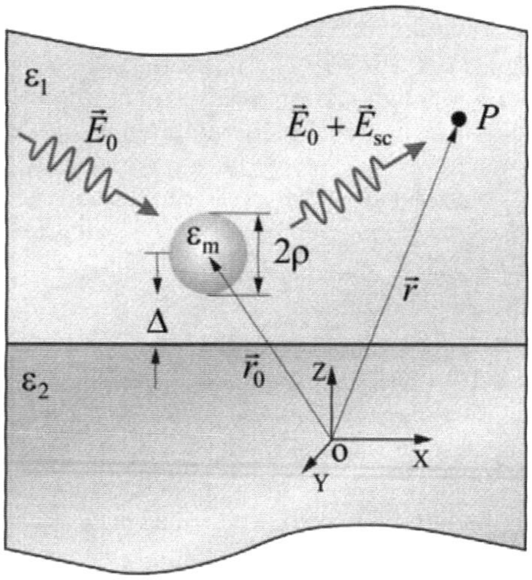

Figure 2. A schematic for optical near-field scattering.

In order to calculate a field distribution near the nanoparticle we need to solve the inhomogeneous Helmholtz equation for an electric field which can be directly obtained from Maxwell's equations [8]:

$$\nabla\times\nabla\times E(r)-k_0^2\varepsilon_1(r)E(r)=\mathrm{i}\omega\mu_0 j_{\mathbf{e}}(r), \tag{3}$$

where k_0 is the wavevector for the incident radiation in vacuum, ω is an angular frequency of the incident radiation and $j_{\mathbf{e}}(r)$ is an induced current density. The solution of Eq. (3) in the formalism of Green's function reads as:

$$\mathbf{E}(\mathbf{r},\omega)=\mathbf{E}_0(\mathbf{r},\omega)+i\omega\mu_0\int_V \ddot{\mathbf{G}}(\mathbf{r},\mathbf{r}_0,\omega)\mathbf{j}_e(\mathbf{r}_0)dV_0, \tag{4}$$

where $\ddot{\mathbf{G}}(\mathbf{r},\mathbf{r_0};\omega)$ is a dyadic Green's function. The current density of an electric dipole with moment $p=\alpha E_0$ located at $r=r_0$ is

$$j_{\mathbf{e}}(r)=-\mathrm{i}\omega p\delta(r-r_0). \tag{5}$$

The Green's function in Eq. (4) includes, in general, these for freely propagating and scattered electromagnetic waves, that is, G_0 and G_{sc}. In particular, the last term considers the scattering from the substrate. With the image method we readily derive

$$\ddot{\mathbf{G}}(\mathbf{r},\mathbf{r_0},\omega)=\left(\ddot{\mathbf{I}}+\frac{1}{k^2}\nabla\nabla\right)\left(\frac{e^{-ik|\mathbf{r}-\mathbf{r_0}|}}{4\pi|\mathbf{r}-\mathbf{r_0}|}-\frac{\varepsilon_2-\varepsilon_1}{\varepsilon_2+\varepsilon_1}\frac{e^{-ik|\mathbf{r}-\mathbf{r}'_0|}}{4\pi|\mathbf{r}-\mathbf{r'_0}|}\right), \tag{6}$$

where r_0' is a vector position of an image dipole.

It is possible to incorporate the influence of the image field into an effective polarizability according to [9]

$$p_{\text{eff}}=\alpha[E_0+E_{\text{image}}]\equiv\alpha_{\text{eff}}E_0, \tag{7}$$

here E_{image} is a local field of the image dipole, which can be written as

$$E_{\text{image}}=\frac{1}{4\pi\varepsilon_0}\left(\frac{3(p_{\text{eff}}(r_0'-r_0))(r'-r_0)}{|r'-r_0|^5}-\frac{p_{\text{eff}}}{|r'-r_0|^3}\right), \tag{8}$$

The polarizability of the small metallic sphere α in Eq. (7) reads as [10]

$$\alpha=4\pi\varepsilon_0\rho^3\begin{pmatrix}\frac{\varepsilon_{\text{m}}-1}{\varepsilon_{\text{m}}+2} & 0 & 0\\ 0 & \frac{\varepsilon_{\text{m}}-1}{\varepsilon_{\text{m}}+2} & 0\\ 0 & 0 & f_{\text{e}}/2\end{pmatrix}, \tag{9}$$

here f_{e} is a complex field enhancement factor depending on tip material and geometry, ε_{m} is a dielectric function of the sphere. Then effective polarizability reads as

$$\alpha_{\text{eff}}^{\|}=\frac{\alpha^{\|}}{1-\alpha^{\|}\text{b}\frac{1}{16\pi\varepsilon_0(\Delta+\rho)^3}}, \tag{10}$$

$$\alpha_{\text{eff}}^{\perp}=\frac{\alpha^{\perp}}{1-\alpha^{\perp}\text{b}\frac{1}{32\pi\varepsilon_0(\Delta+\rho)^3}}, \tag{11}$$

where $b = \frac{\varepsilon_2 - \varepsilon_1}{\varepsilon_2 + \varepsilon_1}$ is a quasi-static Fresnel-reflection coefficient. Changes in the local fields caused by the substrate are equivalent to the modification of tip polarizability. Thus, we can rewrite Eq. (4) as follows:

$$E(\mathbf{r},\omega) = E_0(r,\omega) + \frac{\omega^2}{\varepsilon_0 c^2}\overleftrightarrow{G}_0(r,r_0;\omega)\alpha_{\text{eff}}E_0(r_0,\omega). \tag{12}$$

One of the most common experimental schemes in near-field imaging is the inverted optical configuration, in which the sample is illuminated with the highly focused laser beam with in-plane polarization. Since the antenna mainly interacts with longitudinal lobes we may drop X- and Y-components for the scattered field. In this case we substitute $|\mathbf{r} - \mathbf{r}_0| = \Delta + \rho$, and the component of the dyadic Green's function is written as:

$$G_{iz}^{(0)} = \frac{e^{-ik(\Delta+\rho)}}{4\pi(\Delta+\rho)}\left(\delta_{iz} + i\frac{\delta_{iz}}{k(\Delta+\rho)} + 2\frac{\delta_{iz}}{k^2(\Delta+\rho)^2}\right). \tag{13}$$

Since $k(\Delta+\rho) << 1$ we drop the first and the second terms in this relationship.

By substituting Eq.(10) and Eq.(13) into Eq. (12) and neglecting E_0 compared to the enhanced field, we can readily derive

$$E_z(r,\omega) = \frac{f_e E_{oz}(r_0)e^{-ik(\Delta+\rho)}}{1 - f_e\frac{b\rho^3}{8(\Delta+\rho)^3}}\frac{\rho^3}{(\Delta+\rho)^3}. \tag{14}$$

Figure 3 shows a comparative analysis of the field enhancement obtained with Eq.(14) and that calculated with FDTD method. For this purpose we used the following tip parameters: $\rho = 20$ nm, $f_e = -2.9 + 11.8i$, $\varepsilon_2 = 1.5$ and $\varepsilon_1 = 1$. An excitation wavelength was $\lambda = 632.8$ nm. For numerical simulation the dielectric function for gold was taken from Ref. [11]. In a region where the approximation $k(\Delta+\rho) << 1$ is fulfilled analytical results (black curve) are in a good agreement with the numerical solution of Maxwell's equations (blue dots).

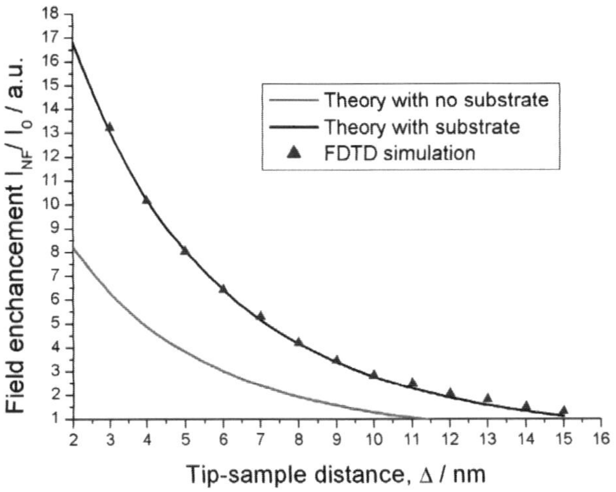

Figure 3. Dependence of the field enhancement vs the tip-substrate distance calculated with Eq.(14) and the FDTD method. A red curve corresponds to the field enhancement when the substrate is absent.

6. Conclusion

In conclusion, we would like to highlight that the field enhancement in the tip-substrate system is caused by two processes: 1) interference between the external and the reflected optical fields; and 2) interaction between the dipole moment of the nanoantenna and the substrate. The first one depends on the material of the substrate only, whereas the second one depends on the parameters of both the tip and the substrate. As a result the field enhancement factor for the tip-substrate system noticeably differs from that for the case when the substrate is not taken into account. This is important for consideration of scattering-type problems in which plasmon-enhanced optical fields are negligible.

Acknowledgments

This study was financially supported by the Russian Foundation for Basic Research (No. 13-02-00758 A).

References

[1] Novotny L 2011 *Physics today* 47-52
[2] Hayazawa N, Yano T and Kawata S 2012 *Journal of Raman Spectroscopy* **43** 1177-1182
[3] Kharintsev S S, Noskov A I, Hoffman G 2007 *Nanotechnology* **18** 315502
[4] Zhu X and Ohtsu M 2000 *World scientific* 205
[5] De Angelis F, Das G, Candeloro P et al. 2009 *Nature Nanotechnology* **5** 67-72
[6] Cancado L G, Jorio A,Ismach A, Joselevich E, Hartschuh A and Novotny L 2009 *Phys. Rev. Lett.* **103** 186101-4
[7] Maximiliano R V, Beams R, Novotny L, Jorio A and Cancado L G 2012 *Phys. Rev. B* **85** 235334-2-8
[8] Novotny L and Hecht B 2006 *Cambridge University Press* 559
[9] Novotny L and Stranick S J 2006 *Annu. Rev. Phys. Chem.* **57** 303-331
[10] Bouhelier A, Beversluis M, Hartschuh A and Novotny L 2003 *Phys. Rev. Lett.* **90** 013903
[11] Johnson P B and Christy R W 1972 *Phys. Rev. B* **6** 4370-4379

DC-pulsed voltage electrochemical method based on duty cycle self-control for producing TERS gold tips

V E Vasilchenko, S S Kharintsev and M Kh Salakhov

Department of Optics and Nanophotonics, Institute of Physics, Kazan Federal University, Kremlevskaya 16, Kazan, 420008, Russia

E-mail: valeriya.vasilchenko@kpfu.ru

Abstract. This paper presents a modified dc-pulsed low voltage electrochemical method in which a duty cycle is self tuned while etching. A higher yield of gold tips suitable for performing tip-enhanced Raman scattering (TERS) measurements is demonstrated. The improvement is caused by the self-control of the etching rate along the full surface of the tip. A capability of the gold tips to enhance a Raman signal is exemplified by TERS spectroscopy of single walled carbon nanotubes bundle, sulfur and vanadium oxide.

1. Introduction

Over the last decade, optical microscopy and spectroscopy have reached the nanoscale through techniques that use optical antennas [1, 2]. Extra information on properties of materials at the nanoscale is obtained by combining scanning probe microscopy and optical spectroscopy. In the case of Raman scattering this approach is often referred to as tip-enhanced Raman spectroscopy (TERS) [3]. In the TERS method a sharp metal tip is illuminated with a laser light in such a way to excite localized surface plasmon waves at the tip apex that concentrate the light energy within of several nanometers. This enables to probe a specimen under study with high spatial resolution, extremely less than the Abbe's diffraction limit, and a giant sensitivity up to several molecules [4]. In general, a tip squeezes the electromagnetic field within a subwavelength volume due to both the lightning rod effect and the plasmonic effect [2]. Based on the latter TERS is a powerful tool for non-destructive chemical analysis of nanometer-sized structures with the spatial resolution of several nanometers [3, 5, 6] and single molecule sensitivity [4] under normal conditions. There is a number of geometrical parameters to be necessarily optimized, namely: curvature radius and cone angle. Tip material is also crucial since it strongly affects plasmon resonances excitation [2]. For the sake of simplicity we used gold wire because of its high performance. In order to produce TERS-active gold tips we utilized a low-cost and reliable electrochemical etching method [6, 7].

In this paper we developed a dc-pulsed low voltage electrochemical method based on a self-adjustable duty cycle for producing TERS active gold tips with a high yield of >90%. Unlike the approach developed in Ref. [9], in our approach *in-situ* cutoff current control is not used. The capacity of gold tips to probe specimen of interest on the nanoscale is demonstrated with near-field Raman measurements of single-walled carbon nanotubes bundle, vanadium oxide and sulfur.

2. Results and discussion

For producing gold tips we used 100 μm gold wire (purity 99.99%) immersed into a solution of fuming hydrochloric acid (HCl, 37%) and ethanol (C_2H_5OH, 96%) in a volume proportion of 1 : 1. A counter-electrode is 10-mm gold ring positioned around the inner bottom-free glass beaker (figure 1(a)) [9]. A sketch of a home-built electrical circuit for etching a gold wire is given in figure 1(b). Microcontroller STM32f415 sets a voltage between the working electrode (WE), which is the gold wire, and the counter electrode (CE) and controls the current cut-off event with a sampling rate of 0.7 μs.

The voltage driven corrosion of pure gold in highly acidic solutions incorporates three processes: 1) oxidation, 2) complexation and 3) diffusion [9]. The complete dissociation of HCl supplies chloride anions Cl^- into the solution which react with a gold wire producing chloroaurate complexes $AuCl_{4-x}[OH]^-_x$ [10]. The most strong etching takes place at the meniscus where Cl^- concentration gradients occur [9]. The presence of chloroaurates prevents the returning of oxidized gold atoms back into the bulk metal. Besides chloroaurates create a double electric layer which blocks an access of Cl^- anions to the surface of the gold wire. One of the ways to attack a passivation layer is mechanical mixing at the meniscus with dc-pulsed voltage. The underlying mechanism is related to change in the meniscus height with applied voltage. For this purpose a voltage square wave is normally utilized. Evidently, such wave parameters as voltage values of V_{down} and V_{up}, a pulse duration of τ, and a duty cycle of $D = \tau/T$ (where T is a period of the square wave) should be optimized. The greater the difference $\Delta V = V_{\text{up}} - V_{\text{down}}$ the greater the extra height of the meniscus $\Delta h(t) = h(V_{\text{up}}) - h(V_{\text{down}})$ [11]. A value of V_{up} is determined in the vicinity of the plateau, as shown in figure 2(b), where the etching current abruptly drops because of the passivation layer. A choice of V_{down} is not limited and it might even be zero to maximize the voltage difference. However, small values of V_{down} lead to vanishing currents and current cutoff event comes true prior to detachment of the lower part of the gold wire. Therefore we choose a value of V_{down} on the plateau which is as far from the value of V_{up} as possible. For the gold wire in hydrochloric acid these values lie in the range of 1.4 V and 2.0 V, as follows from figure 2(b).

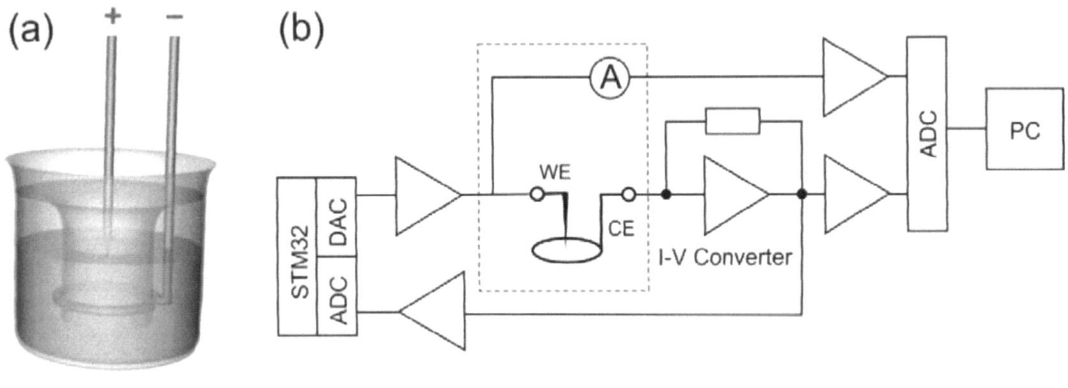

Figure 1. A schematic of the electrochemical cell (a), a layout of an electric circuit for electrochemical etching with a self-tuned duty cycle (b).

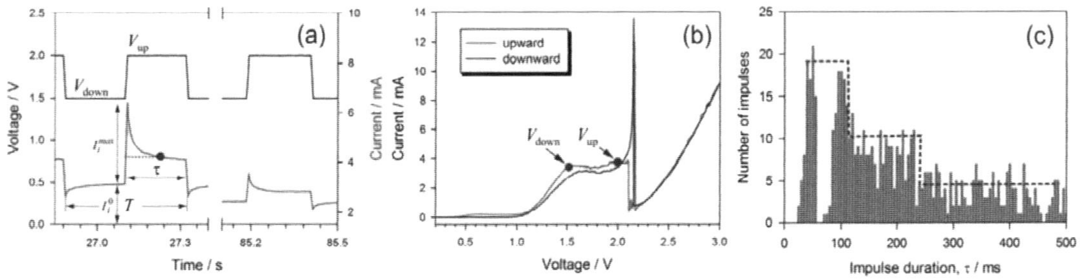

Figure 2. Current kinetics (red curve) and voltage square wave (blue curve) (a), linear sweep voltammetry curves in both directions (b), a histogram of pulse durations (c).

The second parameter of interest is the pulse duration τ or the duty cycle D. Traditionally, this parameter is empirically found from a requirement to improve the reproducibility. However it depends on a number of factors such as background electrolyte, tip material and electrochemical regimes. Besides, the electrochemical etching is assumed to be robust and uniform and, therefore, it makes the tip surface smooth and cone-shaped. Unfortunately an etching rate along the immersed tip surface changes noticeably: convex areas on the surface dissolve stronger than concave ones do. For this purpose we suggest an adaptive electrochemical method in which the duty cycle is self tuned. It means that when the etching current at V_{up} decays down to a threshold value I_{th} the circuit (see figure 1(b)) triggers for V_{down}. As a measure for adaptive control we introduce the parameter η

$$\eta = \frac{I_i - I_i^0}{I_i^{max} - I_i^0}, \quad (1)$$

$$I_{th} = \eta I_i^{max}, \quad (2)$$

where I_i is the current in the *i*-th pulse at V_{up}, I_i^{max} is the maximal current in the *i*-th pulse, I_i^0 is the current at V_{down} (see figure 2(a)). A histogram of pulse durations is given in figure 2(c). From the figure we observe three macro-regions which provide evidence that the gold wire is etched in a non-uniform fashion. It means that the tip shape is not conical, since a radius of the gold wire $r(t)$ diminishes with a variable etching rate. The etching current decay behavior is caused by sliding down the meniscus and thinning the wire due to gravity and corrosion, respectively. The second mechanism is strongly sensitive to small changes in the etching rate when the wire radius approaches to the extent of several nanometers. It means that radius fluctuations at the tip apex are comparable with the ultimate curvature. The conical shape requires uniform etching in correspondence with $r(t) = r_0 - \upsilon t$ (where r_0 is an initial radius of the gold wire, υ is an etching rate). As a result the homogeneous pulse durations distribution has to be observed.

TERS measurements are performed with a scanning probe nanolaboratory NTEGRA SPECTRA™ (NT-MDT, Russia) which includes a confocal scanning laser spectrometer, an optical microscope and a scanning atomic force/shear force microscope. Prepared tips were glued at the end of a quartz tuning fork driven by shear-force control mechanism (see figure 3(a)). The tip apex, shown in figure 3(b), is illuminated with a linearly polarized 632.8 nm wavelength light coming from a He–Ne laser. A 100×oil immersion objective with a high numerical aperture of 1.4 focuses a laser beam down to ~ 250 nm spot. An optical reflection image of the tip raster scanned over the pure glass has demonstrated the presence of longitudinal lobes at the rims of the laser spot (figure 3(c)). All spectra were recorded with an electron multiplying CCD detector (Andor, UK) cooled down to −82°C.

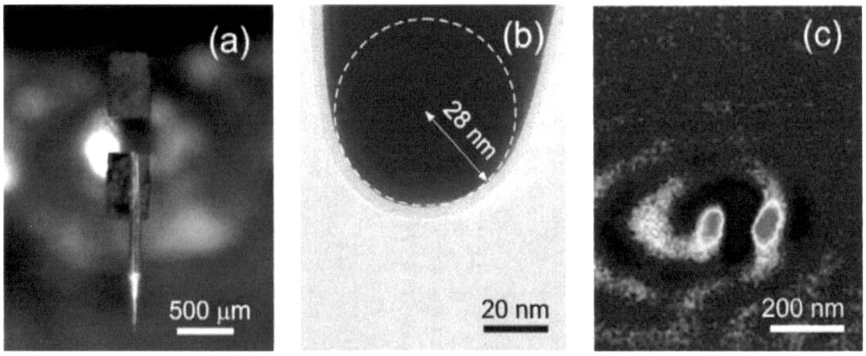

Figure 3. A gold tapered tip glued to a tuning fork operating in the vertical mode (a), high-resolution transmission electron microscopy image of a tip apex (b), confocal optical image of the tip raster scanned over a laser spot (c).

Far- and near-field Raman spectra are exemplified with a single walled carbon nanotubes bundle (figure 4(a)), vanadium oxide (figure 4(b)) and sulfur (figure 4(c)). The results provide evidence that electrochemically etched gold tips, brought to the sample under study with shear force feedback control, operate as hot spots due to the plasmonic effect (average field enhancement factor values exceed 2).

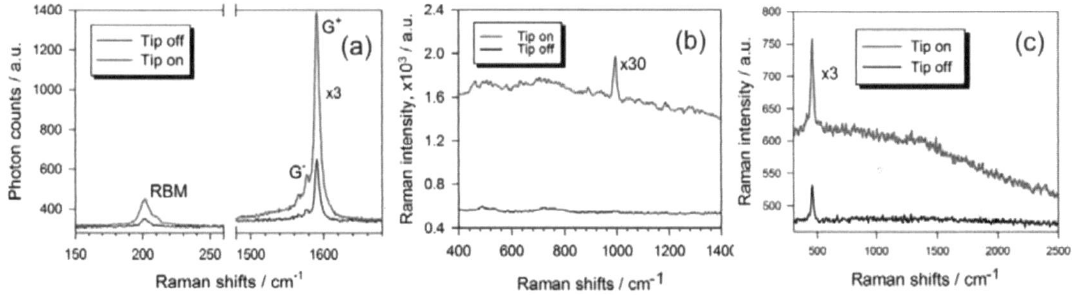

Figure 4. Far (tip-off) and near (tip-on) Raman spectra of single walled carbon nanotubes bundle (a), vanadium oxide (b) and sulfur (c).

3. Concluding remarks

In this paper we have developed a home-built potentiostat which allows one to perform dc-pulsed low voltage etching with self-tuning the duty cycle. In other words, a pulse duration in the voltage square wave is automatically adjusted in the basis of etching rates at the meniscus region. This optimization provides a high yield (>90%) of gold tips with the given parameters when no *in-situ* cut-off current control is used, unlike the approach reported in Ref. [9]. Near-field Raman spectra confirm the presence of the plasmonic effect under illumination of gold tips with a tightly squeezed laser beam.

References

[1] Novotny L Hulst N 2011 *Nature Photonics* **5** 83-90
[2] Novotny L Hecht B 2006 Principles of Nano-optics *Cambridge: Cambridge University Press*
[3] K. Kneipp, M. Moskovits, H. Kneipp 2006 *Topics Applied Physics* **103** 217–240
[4] Zhang W Yeo B S Schmid T Zenobi R 2007 *The Journal of Physical Chemistry C* **111** 1733-1738
[5] Verma P Ichimura T Yano T SaitoY Kawata S 2010 *Laser Photonics Review* **4** 548–561
[6] Stadler J Schmid T Zenobi R 2011 *Nano Lett* **10** 4514–4520

[7] Ren B Picardi B and Pettinger B 2004 *Rev. Sci. Instrum.* **75** 837–41
[8] Billot L Berguiga L de La Chapelle M L Gilbert Y and Bachelot R 2005 *Eur. Phys. J. Appl. Phys.* **31** 139
[9] Kharintsev S S Noskov A I Hoffmann G G Loos J 2011 *Nanotechnology* **22** 025202
[10] Busev A I and Ivanov V M 1973 Analytical Chemistry of Gold *Moscow: Nauka*
[11] Kharintsev S S Hoffmann G G Fishman A I Salakhov M Kh 2013 *Journal of Physics D: Applied Physics* **46** 145501

Journal of Physics: Conference Series **478** (2013) 012017 doi:10.1088/1742-6596/478/1/012017

Electron rest mass and line spectrum of atoms in the photonic crystal medium

R Kh Gainutdinov, M A Khamadeev and M Kh Salakhov

Department of Physics, Kazan Federal University, 18 Kremlevskaya St., Kazan 420008, Russia

E-mail: Renat.Gainutdinov@kpfu.ru

Abstract. We study a recently predicted novel-type quantum electrodynamical effect that takes place in the photonic crystal medium. It consists in the fact that the modification of the interaction of a charged particle with its own radiation field in the photonic crystal medium results in the change in its mass. Consequences and possible applications of this effect are discussed. It is shown that, as a consequence of the effect, spectrum of the light emitted by atoms may be profoundly modified if, instead of being in free space, atoms are enclosed in air voids of a photonic crystal. Spectral lines of atoms in the photonic crystal medium are shifted in frequency, and this shift may be comparable to the atomic transition frequencies in free space.

1. Introduction

There has been a growing interest in the study of photonic crystals (PC) that are a new type of optical material with a periodic dielectric structure. A remarkable feature of photonic crystals is that its periodic dielectric function yields a photonic band structure with photonic band gaps. It has been shown that the gap edge has novel effects on optical behavior of atoms in photonic crystals such as the appearance of photon-atom bound states [1], spectral splitting [2], coherent control of spontaneous emission [3] etc. Strong suppression or enhancement of light emission which take place in PC is expected to modify the Lamb shift in atoms. Earlier studies give rise to controversial results for the Lamb shift. The isotropic dispersion model [1] predicts an anomalous Lamb shift, while the anisotropic model [4] suggests that the Lamb shift should be much smaller than that in vacuum. The calculation of the radiation using the formalism of local density of states (LDOS) resolved the dispute and the prediction of the giant Lamb shift appears to be correct [5]. Thus, the interaction of a bound electron in a photonic crystal with its own radiation field may significantly differs from that in vacuum. However, this may have a significant effect not only on the Lamb shift. In fact, the electron mass with which we usually deal includes the contribution from the interaction with its own radiation field in vacuum. This means that in a photonic crystal the electron mass should be changed. In Ref.[6] it has been shown that in the photonic crystal medium a quantum electrodynamical effect of a new type takes place. Because of the strong modification of the interaction of a charged particle in the PC medium with its own radiation field, the electromagnetic mass in PC differs from its electromagnetic mass in vacuum, which is included into the observable rest mass of the electron. This in turn gives rise to the fact that in the PC medium the electron rest mass changes its value. The effect is of interest both from the fundamental and the practical points of view. In this work we study the effect of the change in the electron rest mass on the linear spectrum of atoms placed in air voids of a photonic crystal and discuss possibilities of the development of novel-type sources with the line spectrum based on this effect.

2. Influence of an environment on the electromagnetic mass of the electron

In describing the QED effects in atoms placed in PCs, one has to take into account the influence of the PC medium on the propagation of virtual photons that come into play in the process of the self-interaction of the atomic electron. Formally, carrying out the mass renormalization for the electron in the PC medium should result in the subtraction of the self-energy of the free electron modified by this medium from the modified self-energy of the bound electron, and this subtraction was used [1, 4, 5, 7–9] in the studies of the Lamb shift in atoms placed in PCs. However, this way of solving the problem leads to missing some important contributions to energy levels from the self-interaction of atomic electrons. In fact, the renormalization theory prescribes to also add the subtracted term m_{em} to the "bare" mass m_0 of the electron in order to arrive at its physical mass $m_e = m_0 + m_{em}$. Here we mean one of the two approaches to renormalization. It has the merit of a clear physical interpretation, but the second approach, the method of counterterms, is the one normally used in quantum field theory. In the second approach the electron mass m_e in the original Lagrangian is regarded as the physical mass. To cancel the contribution from the self-energy of the free electron an extra term is added to the Lagrangian that is called the mass-renormalization counterterm. The problem consists in the fact that the value of the electromagnetic mass of the electron in the PC medium should differ from that in vacuum, and therefore the result of adding this electromagnetic mass to its "bare" mass will not be the physical mass. Obviously, the change of the value of the electron mass δm_{pc} is the difference between the values of the electromagnetic masses in the PC medium m_{em}^{pc} and the electromagnetic mass m_{em} in vacuum

$$\delta m_{pc} = m_{em}^{pc} - m_{em}. \tag{1}$$

Thus, the influence of the PC medium on the interaction of an electron with its own radiation field results in the change in its mass: $m_e \to m_{pc} = m_e + \delta m_{pc}$. Here we assume that the electron is in an air void of the PC.

The PC medium correction δm_{pc} to the electron rest mass is given by the equation [6]

$$\delta m_{pc} = \frac{\alpha}{\mathbf{p}^2 \pi^2} \left[\sum_n \int_{FBZ} \frac{d^3k}{\omega_{\mathbf{k}n}^2} \sum_{\mathbf{G}} |\mathbf{p} \cdot \mathbf{E}_{\mathbf{k}n}(\mathbf{G})|^2 - \int \frac{d^3k}{2\mathbf{k}^2} \sum_{\lambda=1}^{2} |\mathbf{p} \cdot \varepsilon_\lambda(\mathbf{k})|^2 \right] \tag{2}$$

with $\mathbf{E}_{\mathbf{k}n}(\mathbf{G})$ being the coefficients in the plane-wave expansion $\mathbf{E}_{\mathbf{k}n}(\mathbf{r}) = \sum_{\mathbf{G}} \mathbf{E}_{\mathbf{k}n}(\mathbf{G}) e^{i(\mathbf{k}+\mathbf{G})\cdot\mathbf{r}}$ of the Bloch eigenfunctions $\mathbf{E}_{\mathbf{k}n}(\mathbf{r})$. Here $\mathbf{G}$ is the reciprocal lattice vector of the photonic crystal ($\mathbf{G} = N_1\mathbf{b}_1 + N_2\mathbf{b}_2 + N_3\mathbf{b}_3$ where $\mathbf{b}_i$ are primitive basis vectors of a reciprocal lattice). Thus, in contrast to the Lamb shift, the PC medium correction to the electron mass does not depend on the position of the electron in a PC's air void. At the same time, this correction depends on the direction unit vector $\hat{\mathbf{p}} = \mathbf{p}/|\mathbf{p}|$ of the electron momentum.

It should be noted that the mass dependence on orientation of the electron momentum in the PC is not surprising. It is a consequence of anisotropy of the crystal. In solid-state physics, crystal anisotropies result in the fact that the effective mass of an electron depends on direction of the electron momentum with respect to the crystal axes.

3. Linear spectra of the hydrogen atom in a photonic crystal

Let us now discuss perspectives of observation and application of the effect. The change in the electron mass gives rise to the shift of the energy levels of atoms. This shift cannot be classified as a part of the Lamb shift. Let us demonstrate this point by using the example of the atomic hydrogen. In the approximation where the nucleus is assumed to be a point and infinitely massive the energy levels of atomic hydrogen in the photonic crystal medium are given by the solution of the Dirac equation for the energy eigenvalues. For the energy of the atomic state $|n, j, l, m\rangle$, we have

$$\langle \delta m_{pc} \rangle_{n,j,l,m} = \int d^3p \Psi^*_{njlm}(\mathbf{p}) \delta m_{pc}(\hat{\mathbf{p}}) \Psi_{njlm}(\mathbf{p}) \tag{3}$$

with $\Psi_{njlm}(\mathbf{p})$ being the momentum-space hydrogen wave function.

The energy of the hydrogen state in free space $|a\rangle = |n, j, l, m\rangle$ may be written as $E_a = m_e + \epsilon_a$ with $\epsilon_a = -\frac{1}{2}\frac{\alpha^2 m_e}{n^2} + \mathrm{O}(\alpha^4)$. Here the rest energy part of E_a is distinguished. The frequency ω_{ab} of the transition between the state $|a\rangle$ and the state $|b\rangle = |n', j', l', m'\rangle$ is given by

$$\omega_{ab} = \epsilon_a - \epsilon_b. \tag{4}$$

The transition frequency ω_{ab} is equal to $\epsilon_a - \epsilon_b$ because the rest energy contributions are the same for both the states. The situation is dramatically changed in the case when the atom is placed in the PC medium. As we have seen, in this case the rest energy part of the total energy of the bound electron depends on the orbital angular momentum and the angular momentum z component m. As a result, the rest energy parts of the total energies of the states $|a\rangle$ and $|b\rangle$ make a contribution to the transition frequency

$$\omega^{pc}_{ab} = \langle \delta m_{pc} \rangle_a - \langle \delta m_{pc} \rangle_b + \epsilon^{pc}_a - \epsilon^{pc}_b. \tag{5}$$

Thus, in contrast to the free-space case, in the case of the PC medium the rest energy parts of the total energies of the states $|a\rangle$ and $|b\rangle$ make the contribution to the frequency of the transition between these states. Moreover, the difference between $\langle \delta m_{pc} \rangle_a$ and $\langle \delta m_{pc} \rangle_b$ makes a predominant contribution to the PC correction $\delta\omega^{pc}_{ab}$ to the transition frequency

$$\delta\omega^{pc}_{ab} = \left(\langle \delta m_{pc} \rangle_a - \langle \delta m_{pc} \rangle_b\right)\left(1 + \mathrm{O}(\alpha^4)\right) \tag{6}$$

provided $l_a \neq l_b$ and/or $m_a \neq m_b$. Such a surprising appearance of the contribution from the electron rest energy to the atom transition frequencies in the case when the atom is placed in the PC medium gives rise to the fact that the corrections to these frequencies may be very significant. From Eq. (3) it follows that δm_{pc} is independent of the position of the electron in the PC voids but depends on the direction of the electron momentum with respect to the photonic crystal axes. This mass dependence on direction has a significant effect on the structure of the atomic energy levels because it gives rise to the fact that the mean PC medium correction δm_{pc} in states with different orbital angular momenta and/or angular momentum z components are different. This in turn results in the appearance of the term $\langle \delta m_{pc} \rangle_a - \langle \delta m_{pc} \rangle_b$ in the expression (5) for the atomic transition frequency ω_{ab}. Thus, despite that the modification of the interaction of the electron with its own radiation field in the PC medium gives rise to relatively small corrections to the rest electron mass (in our model they are of order $10^{-6} m_e$), it results in the appearance of the term in the expressions for atomic transition frequencies that is absent in the free-space case. This term makes the contribution to ω_{ab} of order $10^{-6} m_e$, while the transition frequency in free space determined by Eq. (4) is given by $\omega_{ab} = \frac{\alpha^2}{2} m_e(\frac{1}{n_b^2} - \frac{1}{n_a^2}) + \mathrm{O}(\alpha^4)$. Thus, the shifts of the energy levels of the atom in the PC medium actually may be comparable to the atomic transition frequency in free space, and for this the modification of the electromagnetic field in this medium need not to be extraordinary. This point may be demonstrated by using the results of the calculations of the mass corrections δm_{pc} to the energy levels of atomic hydrogen in a high-index-contrast photonic crystal. In this way for the transition frequency $2s - 3p, m = 0$

of hydrogen atom in the PC with the $n_{eff} = 3$, where n_{eff} is the refractive index of the corresponding effective homogeneous medium, we get value $\nu_{pc}(2s - 3p, m = 0) = 4.20 \cdot 10^{14}$ Hz, while this frequency in free space is $\nu_{free}(2s - 3p, m = 0) = 3.83 \cdot 10^{14}$ Hz. Thus, the PC medium correction $\delta\nu_{pc} = \nu_{pc} - \nu_{free} = 0.37 \cdot 10^{14}$ Hz is comparable to the transition frequency in free space.

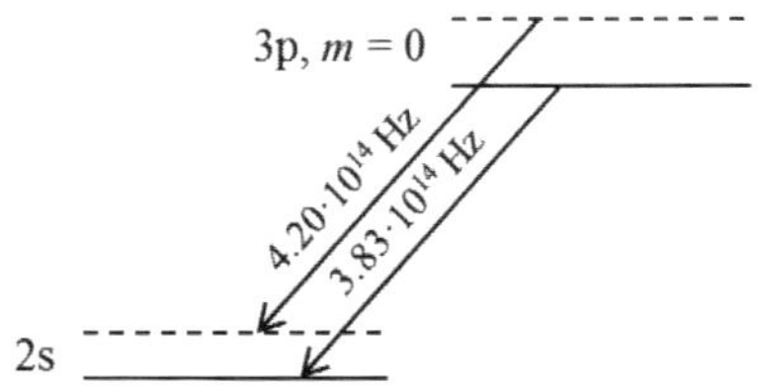

Figure 1. Energy levels of atomic transition frequencies of hydrogen in free space and in the PC medium. The dashed lines denote the energy levels of hydrogen atom placed in a PC.

It must be emphasized that we consider only the atoms placed in a PC's air voids. In this case, in contrast to electrons in ordinary crystals, the atomic electrons are not in contact with the vibrational degrees of freedom of the dielectric host. In this case the influence of the PC medium on atomic electrons is reduced to the influence on the propagation of virtual photons emitted and absorbed by the electrons. This influence can give rise only to the change in the electron mass, and, as a consequence, atoms in the PC should have the same energy-level structure as atoms in free space. Another important point is that the PC medium correction δm_{pc} to the electron mass and hence the atomic frequencies are independent of the position of an atom in a PC's void. As a consequence, spectrum of the light emitted by atom enclosed in PC's voids must be similar to the spectrum of the atoms in free space with the spectral lines shifted in frequency. As we have shown, this shift may be very significant. This gives us the hope that the shift will be observed in near future and the experimental investigations of the effect will give rise to the development of novel-type light sources with the line spectrum. Every spectral line of such sources could be shifted in a wide range by changing properties of a photonic crystal and this is a manifestation of the strong modification of the interaction of a charged particle placed into a PC with its own radiation field. This modification can also have a significant effect on the resonance fluorescence spectrum from emitters placed into a PC. For example, in this case the effect of involving vacuum modes into the strong laser-atom interaction predicted in [11] can take a place. This may be important for quantum-information science implementations since the Mollow triplet in the fluorescence spectrum provides a natural way to spectrally isolate the photons of interest [12].

4. Outlook

We have shown that spectrum of the light emitted by atoms may be profoundly modified if, instead of being in free space, atoms are enclosed in air voids of a PC. Spectral lines of atoms in the PC medium are shifted in frequency, and this shift may be comparable to the atomic transition frequencies in free space. At the same time, we have to keep in mind that this result has been obtained by using the one-loop approximation in describing the electron self-interaction. In solving problems with which the theory of quantum electrodynamics usually deals this approximation obeys a high accuracy. However, because of strong modification of the density of states (DOS) in a photonic crystal from that in free space in some cases the interaction of an atomic electron with its own radiation field may turn out to be effectively strong, and the one-loop approximation may become unapplicable. In this case the change of the electron mass and hence the shift of energy levels of atoms in the photonic crystal medium may be even more significant. Some reason to expect this is given by the effect obtained in Ref. [13] where interaction of an atom with its own radiation field affected by the atomic surrounding was investigated from a general point of view. In Ref. [13] it was shown that for some conditions of the surrounding the interaction of the atom with its own radiation field becomes effectively strong and gives rise to a spectral line splitting. In the case of the photonic crystal medium playing the role of the surroundings, these condition could be realized just because of the singular behavior of the DOS in the PC medium. This provides a way to drive the atomic energy levels. In this way, in particular, light sources with the line spectrum of a new type could be developed.

References

[1] John S and Wang J 1990 *Phys. Rev. Lett.* **64** 2418
[2] John S and Quang T 1994 *Phys. Rev.* A **50** 1764
[3] Quang T, Woldeyohannes M, John S and Agarwal G S 1997 *Phys. Rev. Lett.* **79** 5238
[4] Zhu S Y, Yang Y, Chen H, Zheng H and Zubairy M S 2000 *Phys. Rev. Lett.* **84** 2136
[5] Wang X-H, Kivshar Y S and Gu B-Y 2004 *Phys. Rev. Lett.* **93** 073901
[6] Gainutdinov R Kh, Khamadeev M A and Salakhov M Kh 2012 *Phys. Rev.* A **85** 053836
[7] John S and Wang J 1991 *Phys. Rev.* B **43** 12772
[8] Vats N, John S and Busch K 2002 *Phys. Rev.* A **65** 043808
[9] Li Z Y and Xia Y 2001 *Phys. Rev.* B **63** 121305(R)
[10] Sakoda K 2001 *Optical Properties of Photonic Crystals* (Berlin: Springer)
[11] Gainutdinov R Kh and Mutygullina A A 2009 *Physica Scripta* T **135** 014012
[12] Vamivakas A N, Zhao Y, Lu C-Y and Atatüre M 2002 *Nat. Phys.* **5** 198
[13] Gainutdinov R Kh, Mutygullina A A and Scheid W 2002 *Phys. Lett.* A **306** 1

Nonlinear effect in processes of the interaction of atoms with a resonant laser field

R Kh Gainutdinov and A A Mutygullina

Department of Physics, Kazan Federal University, 18 Kremlevskaya St., Kazan 420008, Russia

E-mail: Renat.Gainutdinov@kpfu.ru

Abstract. We show that for some intensities of a driving laser field not only the laser modes but also the vacuum ones are involved into the strong atom-laser interaction and this has a significant effect on processes of resonance fluorescence.

1. Introduction

The problem of interaction of coherent light with matter is of great interest both from the fundamental point of view as well as for applications. One of the processes that are of fundamental importance to the understanding of this interaction is resonance fluorescence from atoms in an intense laser field. As was predicted by Mollow [1] and confirmed by a number of experiments [2], at relatively high laser intensities the fluorescence spectrum should consist of a symmetric triplet whose side components are separated from the central peak by a distance equal to the generalized Rabi frequency. The strong interaction of an atom with the laser field resonant with an atomic transition is successfully treated within the quantum optics formalism [3, 4]. In describing resonance fluorescence it is usually assumed that only the laser modes are involved into the processes of the strong interaction while other modes including the vacuum ones may be treated as small corrections [5, 6]. However, as has been shown in [7] the Lamb shift in the laser-dressed atoms may be much larger than the Lamb shifts in bare atoms. The aim of the present paper is to show that in some cases not only the atom-laser interaction but also the interaction of the bound electrons with the vacuum modes is strong, and this has a significant effect on the character of the atom-laser interaction.

Let us consider an electron interacting with an atomic nuclear and an intense laser field. It is natural to include the Coulomb field into the free Hamiltonian. This leads us to the Furry picture in which the eigenstates $|m\rangle$ of the Dirac-Coulomb Hamiltonian H_0^F ($H_0^F|m\rangle = E_m|m\rangle$) are used as "free" states, and H_0^F plays the role of a "free" Hamiltonian. When the laser mode frequency ω_L is close to the frequency $\omega_R = \omega_e - \omega_g$ of the transition between two bound states $|e\rangle$ and $|g\rangle$ the atom-laser interaction is strong. In quantum optics this interaction is usually described by using the two-level model and the rotating wave approximation (RWA). In this approximation the laser-dressed states are the eigenstates of the system Hamiltonian H_{RWA} [4] and are given by

$$\begin{aligned} |+,n\rangle &= \cos\theta_n|e,n\rangle + \sin\theta_n|g,n+1\rangle; \\ |-,n\rangle &= -\sin\theta_n|e,n\rangle + \cos\theta_n|g,n+1\rangle. \end{aligned}$$

Here $|e, n\rangle$ ($|g, n\rangle$) denotes the state of the combined laser-atom system containing the atom in the bare state $|e\rangle$ ($|g\rangle$) and n photons in the laser mode, and θ_n is the mixing angle defined by $\tan(2\theta_n) = -\Omega_n/\Delta$ with Ω_n and Δ being the Rabi frequency $\Omega_n = 2g_L\sqrt{n+1}$ and detuning, respectively. The energies of the dressed states are given by $E_{\pm,n} = (n+1/2)\omega_L + (\omega_e - \omega_g)/2 \pm \Omega_R^{(n)}/2$ with $\Omega_R^{(n)}$ being the generalized Rabi frequency $\Omega_R^{(n)} = \sqrt{\Omega_n^2 + \Delta^2}$. As the laser field is assumed to be sufficiently intense, Ω_n, $\Omega_R^{(n)}$ and θ_n are replaced by their semiclassical entities Ω_0, Ω_R and θ in the following discussion. Now we can generalize the Furry picture by adding to the above "free" states the laser-dressed states or, in other words, by adding the interaction that is described by H_{RWA} into the "free" Hamiltonian in the Furry picture $H_0^F \to H_0^{\mathcal{D}}$. In this case the "free" Green operator $G_0^F(z) = (z - H_0^F)$ is replaced by the operator $G_0^{\mathcal{D}}(z) = (z - H_0^{\mathcal{D}})^{-1}$, and the eigenvectors of $H_0^{\mathcal{D}}$ ($H_0^{\mathcal{D}}|m'\rangle = E_{m'}|m'\rangle$, $H_0^{\mathcal{D}}|\pm, n; m\rangle = (E_{\pm,n} + E_m)|\pm, n; m\rangle$, $\langle m'|\pm, n; m\rangle = 0$), can be used as basic vectors in the Fock space.

2. The Generalized Quantum Dynamics

In our investigation of the problem we do not assume a priori that the QED corrections are small. As will be shown below, a natural way to solve the problem in this case is provided by the generalized dynamical equation (GDE), which in [8] has been derived as a direct consequence of the first principles of quantum physics. Being equivalent to the Schrödinger equation in the case when the interaction in a quantum system is instantaneous, the GDE allows one to extend quantum dynamics to the case of nonlocal-in-time interactions. Generalized quantum dynamics (GQD) developed in this way provides a new insight into the many problems in quantum physics [9–13]. In the GQD a history of a physical process is represented by some version of the time evolution of the system associated with completely specified instants of the beginning and end of the interaction in the system [8]:

$$U(t, t_0) = 1 + \int_{t_0}^{t} dt_2 \int_{t_0}^{t_2} dt_1 \tilde{S}(t_2, t_1) \quad (1)$$

where $\tilde{S}(t_2, t_1)$ describes a contribution to the evolution operator $U(t, t_0)$ from the process in which the interaction begins at time t_1 and ends at time t_2. The GDE, which has the form

$$\tilde{S}(t_2, t_1) = \int_{t_1}^{t_2} dt_4 \int_{t_1}^{t_4} dt_3 \frac{(t_4 - t_3)}{t_2 - t_1} \tilde{S}(t_2, t_4)\tilde{S}(t_3, t_1),$$

allows one to obtain $\tilde{S}(t_2, t_1)$ for any t_1 and t_2, if the operator $\tilde{S}(t_2', t_1')$ corresponding to infinitesimal duration times $\tau = t_2' - t_1'$ of interaction are known. Most of the contribution to the evolution operator in the limit $\tau \to 0$ comes from processes associated with the fundamental interaction in the system under study, which can be parameterized as an interaction operator $H_{int}(t_2, t_1)$ [8]. This gives rise to the following boundary condition for $\tilde{S}(t_2, t_1)$:

$$\tilde{S}(t_2, t_1) \underset{t_2 \to t_1}{\longrightarrow} H_{\mathrm{int}}(t_2, t_1).$$

In the Schrödinger picture the evolution operator (1) can be represented in the form [9]

$$U_s(t, 0) = \frac{i}{2\pi} \int_{-\infty}^{\infty} dz \exp(-izt) G(z),$$

with $G(z) = G_0(z) + G_0(z)T(z)G_0$. Here $G_0(z) \equiv (z - H_0)^{-1}$ is the free Green operator and the operator $T(z)$ is defined as

$$T(z) = i \int_0^{\infty} d\tau \exp(-iz\tau)\tilde{T}(\tau),$$

where $\tilde{T}(t_2-t_1)=\exp(-iH_0t_2)\tilde{S}(t_2,t_1)\exp(iH_0t_1)$. In terms of the operator $T(z)$ the GDE and the corresponding boundary condition take the form [8]

$$\frac{dT(z)}{dz}=-T(z)(G_0(z))^2T(z), \tag{2}$$

$$T(z)\underset{|z|\to\infty}{\longrightarrow}B(z)\equiv i\int_0^\infty d\tau\exp(iz\tau)\tilde{B}(\tau), \tag{3}$$

with $\tilde{B}(\tau)=\exp(-iH_0t_2)H_{int}(t_2,t_1)\exp(iH_0t_1)$. The contribution to the Green operator $G(z)$, which comes from the processes associated with the self-interaction of the particles, has the same structure as the free Green operator $G_0(z)$. For this reason it is natural to replace $G_0(z)$ by the propagator $\tilde{G}_0(z)$, which describes the evolution of particles interacting only with the vacuum and hence has the structure

$$\tilde{G}_0(z)=(z-H_0-C(z))^{-1},$$

where the operator $C(z)$ is determined by the equation $C(z)|n\rangle=C_n(z)|n\rangle$ with $|n\rangle$ being the eigenvectors of the free Hamiltonian $H_0|n\rangle=E_n|n\rangle$ and the condition $z-E_n-C_n(z)=0$ determines the physical masses of the particles. Correspondingly, the operator $T(z)$ should be replaced by the operator $M(z)$ which describes the evolution of particles interacting not only with the vacuum. These operators are related as follows:

$$G_0(z)+G_0(z)T(z)G_0(z)=\tilde{G}_0(z)+\tilde{G}_0(z)M(z)\tilde{G}_0(z).$$

By using this relation, one can rewrite (2) in terms of $M(z)$ and $C_n(z)$, satisfying the boundary conditions

$$M(z)\underset{|z|\to\infty}{\longrightarrow}B_r(z),\quad C_n(z)\underset{|z|\to\infty}{\longrightarrow}\langle n|B_\delta(z)|n\rangle.$$

Here $B_r(z)$ is the part of the interaction operator describing the proper interaction between the particles, and the "singular" part $B_\delta(z)$ describes their self-interaction. This trick was firstly used in [14]. The features of the above representation of the GDE is very useful for solving the bound-state problem in QED. In order to obtain the self-energy corrections to the bound states of an atom dressed by the resonant laser field one should start from the operator $G_0^{\mathcal{D}}(z)$ that in the generalized Furry picture plays the role of the "free" Green operator. In this case the equation $z-E_m-C_m(z)=0$ determines the self-energy corrections to the atomic energy levels.

3. Vacuum modes and the strong atom-laser interaction

The resonant laser field has an effect on the self-interaction of the atom only in the states $|e\rangle$ and $|g\rangle$. For this reason in solving the problem we can include the self-energy corrections to the energies of other states from the every beginning. In this case only the self-energy functions $C_{\pm,n}(z)$ do not equal to zero, and the above reduction of Eq. (2) gives rise to the equations

$$\frac{dC_{\pm,n}(z)}{dz}=-\langle\pm,n|M(z)(\tilde{G}_0(z))^2M(z)|\pm,n\rangle, \tag{4}$$

$$\begin{aligned}\frac{dM(z)}{dz}=&-(1-P(z)\tilde{G}_0(z))M(z)(\tilde{G}_0(z))^2M(z)(1-\tilde{G}_0(z)P(z))+\\&+P(z)\tilde{G}_0(z)M(z)(\tilde{G}_0(z))^2M(z)\tilde{G}_0(z)P(z),\end{aligned} \tag{5}$$

where

$$P(z)=\sum_{n,m}(P_m^+M(z)P_m^-+P_m^-M(z)P_m^+),$$

$$P_m^\pm=|\pm,n;m\rangle\langle\pm,n;m|,$$

$$\tilde{G}_0(z) = (z - H_0^{\mathcal{D}} - C(z))^{-1},$$

and $C(z)|\pm, n, m\rangle = C_{\pm,n}(z - E_m)|\pm, n, m\rangle$. In this case the interaction operator in the boundary condition (3) for these equations is of the form $B(z) = H_I + B_\delta^{\mathcal{D}}(z)$, where $H_I = e\int d^3x : A_\mu(0,\mathbf{x})\bar{\psi}(0,\mathbf{x})\gamma^\mu\psi(0,\mathbf{x}) :$, $\psi(t,\mathbf{x})$ is the Dirac field in the Furry picture, $A^\mu(x)$ is the electromagnetic field, the symbol : : denotes the normal ordering, and $B_\delta^{\mathcal{D}}(z)$ is a nonlocal operator describing the self-interaction of electrons in the generalized Furry picture. In some cases Eqs. (4) and (5) can be solved iteratively. Putting $M(z) = H_I$ on the right-hand part of Eqs. (4) and (5), for $\langle\pm, n|M(z)|\mp, n\rangle \equiv \langle\pm, n|\Sigma(z)|\mp, n\rangle$ and $C_{\pm,n}(z) \equiv \langle\pm, n|\Sigma(z)|\pm, n\rangle$, we get the equation

$$\frac{d\langle\psi_2|\Sigma^{(0)}(z)|\psi_1\rangle}{dz} = -\langle\psi_2|H_I(G_0^{\mathcal{D}}(z))^2 H_I|\psi_1\rangle. \tag{6}$$

In calculating with the accuracy up to order $o\{\Lambda^2\}$, where $\Lambda = \Omega_R/\alpha^2 Z^2 m_e$ (in general this parameter is extremely small), $G_0^{\mathcal{D}}$ in Eq. (6) can be replaced by $G_0^F(z)$ and the operator $B_\delta^{\mathcal{D}}$ is a simple redistribution of $B_e(z)$ and $B_g(z)$ being the eigenvalues of the operator $B_\delta^F(z)$ describing the self-interaction of the bare atoms ($B_\delta^F(z)|e\rangle = B_{\delta,e}(z)|e\rangle$, $B_\delta^F(z)|g\rangle = B_{\delta,g}(z)|g\rangle$). In this approximation the solution of Eq.(6) is

$$\begin{aligned} \langle +, n|\Sigma^{(0)}(z)|+, n\rangle &= \cos^2\theta C_e(z') + \sin^2\theta C_g(z''), \\ \langle -, n|\Sigma^{(0)}(z)|-, n\rangle &= \sin^2\theta C_e(z') + \cos^2\theta C_g(z''), \end{aligned} \tag{7}$$

$$\langle\pm, n|M^{(0)}(z)|\mp, n\rangle = \frac{1}{2}\sin 2\theta(C_e(z') - C_g(z'')), \tag{8}$$

where $z' = z - n\omega_L$, $z'' = z - (n+1)\omega_L$, and

$$C_i(z) = \lim_{\mu\to\infty}\langle i|H_I(G_0^F(z) - G_0^F(\mu))H_I|i\rangle + B_{\delta,i}(\mu), \quad i = e, g \tag{9}$$

Here μ plays the role of the regularization parameter, and $\langle i|B_\delta^F(\mu)|i\rangle$ manifests itself as a sum of μ-dependent counterterms. Thus, the leading order solution of Eqs. (4) and (5) is

$$\begin{aligned} C_{\pm,n}^{(0)}(z) &\equiv \langle\pm, n|\Sigma^{(0)}(z)|\pm, n\rangle, \\ M^{(0)}(z) &= H_I + \mathcal{L}(z), \\ \mathcal{L}(z) &= \sum_{n,m} P_m^+(\Sigma^{(0)}(z)P_m^- + P_m^-\Sigma^{(0)}(z)P_m^+). \end{aligned} \tag{10}$$

with $\langle\pm, n|\Sigma^{(0)}(z)|\pm, n\rangle$ being defined by Eqs. (7). The next iteration yields

$$C_{\pm,n}^{(1)}(z) = C_{\pm,n}^{(0)}(z) + \langle\pm, n|M^{(1)}(z)\tilde{G}_0^{(0)}(z)M^{(1)}(z) - H_I\tilde{G}_0^{(0)}(z)H_I|\pm, n\rangle,$$

$$M^{(1)}(z) = M^{(0)}(z) + H_I\tilde{G}_0^{(0)}(z)\mathcal{L}(z) + \mathcal{L}(z)\tilde{G}_0^{(0)}(z)H_I, \tag{11}$$

where $\tilde{G}_0^{(0)}(z) = z - H_0^{\mathcal{D}} - C^{(0)}$. As it follows from these equations, the correction to the leading order solution is of the relative order $O(\Lambda_\theta)$ where $\Lambda_\theta \equiv -\sin(2\theta)(\delta E_g - \delta E_e)/2\Omega_R$, and the iterative scheme yields the expansion of the solution of Eqs. (4) and (5) in powers of this parameter. In the case $\Omega \gg \Gamma_e$ the decay law may be close to an exponential one and for describing the fluorescence spectrum one can use the standard master equations even if the interaction of an atom with the vacuum modes is effectively strong. In this way one can use the simple rules derived in Ref. [15] for evaluating the positions, the heights and the widths of the various components of the fluorescence spectrum. In particular, these rules relate the widths Γ_{ij} of the transitions from $|i,n\rangle$ to $|j,n-1\rangle$ (here and below we use the notation $|1,n\rangle \equiv |-,n\rangle$ and $|2,n\rangle \equiv |+,n\rangle$) with the widths of the central and the sideband components of the fluorescence spectrum respectively $L_{ii} = \Gamma_{ii} + \Gamma_{jj}$, $L_{ij} = L_{ji} = \frac{1}{2}(\Gamma_i + \Gamma_j) - \mathbf{d}^-_{ii} \cdot \mathbf{d}^+_{jj}$, where

$$\Gamma_{ij} = |\mathbf{d}^+_{ij}|^2, \quad \mathbf{d}^\pm_{ij} = \langle i,n|\mathbf{D}|j,n\mp 1\rangle, \tag{12}$$

and $\mathbf{D}$ is the dipole momentum operator. Here the factor $a = \sqrt{\frac{\omega_R}{3\pi}}$ is reincluded in the definition of $\mathbf{D}$ in order to get a simple expression for Γ. In deriving these formulas it is assumed that with the accuracy of order $O(\alpha)$ the transitions between the states $|i,n\rangle$ and $|j,n-1;\mathbf{k},\varepsilon_\lambda\rangle$ are described by the corresponding matrix elements of the interaction Hamiltonian H_I in the dipole approximation $H_I = -a^{-1}\mathbf{D}\cdot\varepsilon_\lambda$ where $\mathbf{k}$ and ε_λ are the photon momentum and polarization respectively. However, taking into account vacuum modes implies that the coupling of the laser-dressed atom to the radiation field is described by the corresponding matrix elements of the operator $M(z)$, and hence the dipole momentum in Eq.(12) is defined as $\langle j,n-1;\mathbf{k},\varepsilon_\lambda|M^{(1)}(E_i)|i,n\rangle = -a^{-1}\langle j,n-1|\mathbf{D}_v\cdot\varepsilon_\lambda|i,n\rangle$. From Eq.(10) it follows that $M(z)$ is reduced to H_I only at leading order and the higher order corrections can be significant.

4. Outlook

In summary, we have shown that the vacuum modes may be involved into the strong interaction of an atom with a resonant laser field. As it follows from Eq.(8), the influence of this nonlinear effect on resonance fluorescence depends on the self-energy functions $C_e(z)$ and $C_g(z)$ of the relevant states of the bare atom. In free space the dependence of $C_i(z)$ on z in the energy interval between $E_{-,n}$ and $E_{+,n}$ is negligible. This allows one to diagonalize $\Sigma^{(0)}(z)$ in Eq.(10) and to get the solution without further iterations. The solution obtained in this way is reduced to a change of the mixing angle in the definition of the dressed states θ_n. The situation is cardinally changed in the case when the atom is placed in a photonic crystal (PC) being periodic nanostructures. Because of the strong modification of the interaction of an electron in a PC with the radiation field which can even results in the electron mass changes its value [16], the corresponding self-energy functions are significantly changed in the above energy interval and hence in this case the effect can be significant.

References

[1] Mollow B R 1969 *Phys. Rev.* **188** 1969
[2] Wu F Y, Grove R E and Ezekiel S 1975 *Phys. Rev. Lett.* **35** 1426
Shuda F, Stroud Jr C R and Hercher M 1974 *J. Phys.* B **198** L7
Grove R E, Wu F Y and Ezekiel S 1977 *Phys. Rev.* A **15** 227
[3] Mandel L and Wolf E 1995 *Optical Coherence and Quantum Optics* (Cambridge: Cambridge University Press)
[4] Scully M and Zubairy M 1997 *Quantum Optics* (Cambridge: Cambridge University Press)
[5] Jentschura U D and Keitel Ch H 2004 *Ann. Phys.* **310** 1
[6] Jentschura U D, Evers J, Haas M and Keitel Ch H 2003 *Phys. Rev. Lett.* **89** 253601
[7] Gainutdinov R Kh and Mutygullina A A 2009 *Physica Scripta* T **135** 014012
[8] Gainutdinov R Kh 1999 *J. Phys.* A **32** 5657
[9] Gainutdinov R Kh and Mutygullina A A 2002 *Phys. Rev.* C **66** 014006
[10] Gainutdinov R Kh, Mutygullina A A and Scheid W 2002 *Phys. Lett.* A **306** 1
[11] Gainutdinov R Kh, Moutygoullina A A 1997 *Physics of Atomic Nuclei* **60** 841
[12] Gainutdinov R Kh, Moutygoullina A A 1999 *Physics of Atomic Nuclei* **62** 1905
[13] Gainutdinov R Kh, Moutygoullina A A 2002 *Physics of Atomic Nuclei* **65** 1421
[14] Gainutdinov R Kh 1989 *J. Phys.* A **22** 269
[15] Cohen-Tannoudji C and Reynaud S 1976 *J. Phys.* B **10** 345
[16] Gainutdinov R Kh, Khamadeev M A and Salakhov M Kh 2012 *Phys. Rev.* A **85** 053836.

Journal of Physics: Conference Series **478** (2013) 012019 doi:10.1088/1742-6596/478/1/012019

Unstable vacuum and spectrum of atoms with a superheavy nucleus

R Kh Gainutdinov, A A Mutygullina, M A Khamadeev, A S Petrova

Kazan Federal University, 18 Kremlyovskaya St., Kazan 420008, Republic of Tatarstan, Russian Federation

E-mail: Renat.Gainutdinov@kpfu.ru

Abstract. We show that the light spectra of atoms with a superheavy nucleus may be important for our understanding of the processes in supercritical fields. The shape is obtained of a spectral line emitted by an atom with a superheavy nucleus. We study the connection of the light part of the spectrum with the parameters characterizing the nucleus decay.

1. Introduction

Optical spectroscopy has played an important part in the development of physics as a whole. Observation of the Balmer series in the Hydrogen emission spectrum inspired the Bohr atomic model and quantum mechanics. The discovery of the Lamb shift led to the creation of quantum electrodynamics (QED). Modern laser spectroscopy allows one to reach a precision of 15 decimal digits and to register radiation of very low intensity, up to one photon, providing a powerful tool for investigation and tests of fundamental physics laws [1]. So, the spectroscopy of mounic hydrogen has led to the well-known proton radius puzzle [2] and made scientists review the fundamentals of QED. Another questionable problem, which provokes interest during last several decades, is the spontaneous real electron-positron e^-e^+ pair creation in heavy ion collisions [3]. It is expected that at certain bombarding energies in the collisions of the heaviest ions (such as for example Uranium ^{238}U or Curium ^{248}Cm) giant nuclei with the supercritical charge $Z = 188$, or more, may be formed. This should lead to the spontaneous real electron-positron e^-e^+ pair creation from the vacuum and, as a consequence, to the appearance of additional narrow lines with completely defined position and shape in the positron energy spectrum. The existence of these lines has not been unambiguously established experimentally yet [4,5]. The key problem is that there are several mechanisms of positron production during a heavy ion collision, such as induced positron creation and internal pair conversion [6], and it is impossible to distinguish experimentally between the contributions from these mechanisms. Besides, there are still doubts that giant nuclei live long enough for the vacuum decay to occur. Thus, positron spectra provide not enough data and an additional independent source of information is needed to resolve the problem. Optical spectra of heavy ion collisions may serve this source, since their structure is vastly affected by the inner parameter of the giant nucleus. In this paper we derive the form of the spectrum of the optical radiation emitted by an atom with a supercritical nucleus. We show how it is affected by the nuclear characteristics mentioned above. The investigation of optical spectra is universal and may be applied for the study of any other ions, to determine their physical parameters.

2. Methods

In our investigation of the problem we use the generalized dynamical equation (GDE), which in [7] has been derived as a direct consequence of the first principles of quantum physics. Being equivalent to the Schrödinger equation in the case when the interaction in a quantum system is instantaneous, GDE allows one to extend quantum dynamics to the case of nonlocal-in-time interactions. Generalized quantum dynamics (GQD) developed in this way provides a new insight into many problems in quantum physics [8-13]. In GQD a history of a physical process is represented by some version of the time evolution of the system associated with completely specified instants of the beginning and end of the interaction in the system [7]:

$$U(t,t_0) = 1 + \int_{t_0}^{t} dt_2 \int_{t_0}^{t_2} dt_1 \tilde{S}(t_2,t_1), \tag{1}$$

where $\tilde{S}(t_2,t_1)$ describes a contribution to the evolution operator $U(t,t_0)$ from the process in which the interaction begins at time t_1 and ends at time t_2. GDE allows one to obtain $\tilde{S}(t_2,t_1)$ for any t_1 and t_2, if the operators $\tilde{S}(t_2',t_1')$ corresponding to infinitesimal duration times $\tau = t_2' - t_1'$ of interaction are known. Most of the contribution to the evolution operator in the limit $\tau \to 0$ comes from processes associated with the fundamental interaction in the system under study, which can be parameterized as an interaction operator $H_{\text{int}}(t_2,t_1)$ [7]. This rise to the following boundary condition for $\tilde{S}(t_2,t_1)$:

$$\tilde{S}(t_2,t_1) \xrightarrow[t_2 \to t_1]{} H_{\text{int}}(t_2,t_1).$$

In the Schrödinger picture the evolution operator can be represented in the form [9]

$$U_s(t,0) = \frac{1}{2\pi} \int_{-\infty}^{\infty} dx \exp(-iz\tau) G(z), \tag{2}$$

with $G(z) = G_0(z) + G_0(z) T(z) G_0(z)$, where $G_0(z) = (z - H_0)^{-1}$, H_0 is the free Hamiltonian, and the operator $T(z)$ is defined as

$$T(z) = i \int d\tau \exp(iz\tau) \exp(iH_0 t_2) \tilde{S}(t_2,t_1) \exp(iH_0 t_1), \tag{3}$$

where $\tau = t_2 - t_1$. In terms of the operator $T(z)$, GDE takes the form

$$\frac{dT(z)}{dz} = -T(z)\left(G_0(z)\right)T(z), \tag{4}$$

which is used with the boundary condition $T(z) \xrightarrow[|z| \to \infty]{} B(z)$, where

$$B(z) = i \int d\tau \exp(iz\tau) \exp(iH_0 t_2) H_{\text{int}}(t_2,t_1) \exp(-iH_0 t_1). \tag{5}$$

The contribution to the Green operator $G(z)$, which comes from the processes associated with the self-interaction of particles, has the same structure as the free Green operator $G_0(z)$. For this reason it is natural to replace $G_0(z)$ by the propagator $\tilde{G}_0(z)$, which describes the evolution of particles interacting only with the vacuum and hence has the structure

$$\langle m' | \tilde{G}_0(z) | m \rangle = \langle m' | m \rangle \left(z - E_m - C_m(z)\right)^{-1}, \tag{6}$$

where $|m\rangle$ are the eigenvectors of the free Hamiltonian ($H_0|n\rangle = E_n|n\rangle$). Correspondingly, the operator $T(z)$ should be replaced by the operator $M(z)$, which describes the evolution of particles interacting not only with the vacuum. These operators are related as follows:

$$G(z) = G_0(z) + G_0(z)T(z)G_0(z) = \tilde{G}_0(z) + \tilde{G}_0(z)M(z)\tilde{G}_0(z). \tag{7}$$

The self-energy function $C_m(z)$ describes the self-interaction of particles in the state $|m\rangle$, and the condition $z - E_m - C_m(z) = 0$ determines the physical masses of the particles. Taking into account equation (7), one can rewrite (4) in terms of $M(z)$ and $C_m(z)$, with the boundary conditions $M(z)\xrightarrow[|z|\to\infty]{} B_r(z)$, $C_m(z)\xrightarrow[|z|\to\infty]{} B_\delta(z)$. Here $B_r(z)$ is the part of the interaction operator describing the proper interaction between particles, and $B_\delta(z)$ describes their self-interaction: $B(z) = B_r(z) + B_\delta(z)$. This trick was firstly used in [14].

3. Supercritical atoms

The Dirac equation with the Coulomb potential $V(\mathbf{r}) = -Z\alpha\mathbf{r}^{-1}$ (α is the fine-structure constant) allows one to find atomic energy levels for any nucleus charge number $Z \le 137$ in the case of a point nucleus and $Z \le Z_{cr} \approx 173$ for an extended nucleus [3]. If the charge of the nucleus becomes larger than Z_{cr}, the binding energy E_{1s} of the lowest electron becomes smaller than the electron-positron pair rest energy $E_{1s} \le -2m_e$. The electron K-shell joins the positron energy continuum. If it is empty, an additional vacancy in the lower continuum is created, which may be spontaneously occupied by an electron from this continuum leaving a hole in it, what is equivalent to the positron emission. This process is called the vacuum decay in supercritical fields [3]. It results in the appearance of additional narrow lines in the positron kinetic energy spectrum. The standard approach that uses the Dirac equation predicts the widths of the lines to be several keV, what corresponds to the decay time $\tau_d = \hbar/\Gamma \sim 10^{-19}$ s [3], while the lifetime of giant nuclei formed during the collision is predicted to be $\tau_N \sim 10^{-20}$ s [6], what is obviously much shorter than τ_d. The narrow line structures were not confirmed by the last experiments on heavy ion collisions [4,5]. Thus, the situation remains unclear.

In general case, an unstable state is not an eigenstate of any Hamiltonian and is described by an energy distribution $\omega_i(E)$. The vector of an unstable state may be written in the form $|\phi_i\rangle = \int a_i(E)|E,\psi_i\rangle dE$, where $|E,\psi_i\rangle$ is the eigenvector of the energy operator corresponding to the energy E. $\omega_i(E) = |a_i(E)|^2$ determines the decay law of the unstable state unambiguously.

In description of electron motion in the field of a nucleus the Furry picture is used, in which the free Green operator is constructed on the basis of the eigenvectors of Hamiltonian $\tilde{H}_0$ including the Coulomb interaction of the electrons with the nucleus. In our case these are the states of the critical Hamiltonian $H_{cr} = i\gamma_v(\partial/\partial x_v) + m_e - Z_{cr}\alpha\mathbf{r}^{-1}$, $H_{cr}|\psi_n^{(cr)}\rangle = E_n^0|\psi_n^{(cr)}\rangle$. The "free" Green operator (6) in this basis looks as follows:

$$\tilde{G}_0^{(cr)}(z) = \sum_n \frac{|\psi_n^{(cr)}\rangle\langle\psi_n^{(cr)}|}{z - E_n^{(0)} - C_n(z)}, \tag{8}$$

where $C_n(z)$ is the self-energy function of the state $|\psi_n^{(cr)}\rangle$, characterizing the interaction of the supercritical atom with the vacuum. The energy distribution $\omega_{1s}(E)$ is defined as follows:

$$\omega_{1s}(E) = |a_{1s}(E)|^2 = (2\pi)^{-2}\left(z - E_{1s} - C_{1s}(z)\right)^{-2}. \tag{9}$$

It may be found that in the first order in $\lambda = Z'/Z_{cr}$

$$\omega_{1s}^{(2)} = \left|a_{1s}^{(2)}(E)\right|^2 = \frac{1}{2\pi}\frac{\Gamma}{\left[z - E_{1s}^{(0)} - \Delta E_{1s} - F(z)\right]^2 + \Gamma^2/4}, \quad (10)$$

where $\Gamma = 2\pi\left\langle\psi_{1s}^{(cr)}\middle|V(Z')\middle|\phi_E^{(cr)}\right\rangle$, $\Delta E_{1s} = <\psi_{1s}^{(cr)}|V(Z')|\psi_{1s}^{(cr)}>$ and $\left|\phi_E^{(cr)}\right\rangle$ are the eigenstates of H_{cr}, belonging to the negative energy continuum. Expression (10) has the Lorentz form with the width of several keV, what coincides with the results obtained in [3]. But in the next order in λ energy distribution $\omega_{1s}^{(4)}$ is more complicated and can be obtained only within the framework of GQD. In any case, solving the problem by means of the introduction of the overcritical Coulomb potential leads to the width of the energy distribution of several keV. Note, that this quantity characterizes a one-electron state, rather than the vacuum state.

To describe exactly the vacuum decay we need to find the energy of the interaction of a bare superheavy nucleus with the vacuum $C_N(z)$. It can be shown from the dimensional analysis of the GDE for the function $C_N(z)$ that it is of the order of the electron rest energy $C_N(z) \sim m_e$, what corresponds to the decay time magnitude $\tau_d \sim 10^{-21}$ s. This result is in better correspondence with the nuclear lifetime prediction, but leads to a considerable broadening of the spontaneously created positron lines in comparison with the results given in [3].

4. Light spectrum of supercritical atoms

The emission of photons by a superheavy atom is caused by the transition of electrons between the energy levels with high principal quantum numbers n and m. Let us consider the following evolution process of an atom with a supercritical double-uranium nucleus $Z = 188$. An atomic electron transits from an excited state $|N_i\rangle$ to any lower state $\left|N_f;\bar{k},\varepsilon_\lambda\right\rangle$ and emits an optical photon with the energy ω. Here N_i and N_f stand for the entire set of discrete and continuous variables that characterize respectively the initial and the final states of the system in full (for example, $N_i = np_{1/2}$ and $N_f = ms_{1/2}$), $\bar{k}$ and ε_λ are momentum and polarization vector of the photon. After the transition the atomic nucleus splits into two Uranium nuclei, each having the energy E_U. The natural broadening of spectral line profiles is determined by the formula [15]:

$$P(\omega) = \frac{dW_i(\omega)}{d\omega} = A'\omega\sum_\lambda\int d\Omega_k\left|\left\langle N_f;\bar{k},\varepsilon_\lambda\middle|U(t,0)\middle|N_i\right\rangle\right|^2, \quad (11)$$

where A' is a normalization factor. The full Green operator corresponding to the above process is:

$$\left\langle N_f;k,\varepsilon_\lambda\middle|G(z)\middle|N_i\right\rangle = \sum_n\sum_{n'}\frac{\left\langle N_f;\bar{k},\varepsilon_\lambda\middle|\tilde{G}_0(z)\middle|n';\bar{k},\varepsilon_\lambda\right\rangle\left\langle n';\bar{k},\varepsilon_\lambda\middle|M(z)\middle|n\right\rangle\left\langle n\middle|\tilde{G}_0\middle|N_i\right\rangle}{z - 2E_U - \omega}, \quad (12)$$

where $\left\langle N_i\middle|\tilde{G}_0\middle|N_i\right\rangle = \left[z - E_N - C(z)\right]^{-1}$ and $\left\langle N_f;\bar{k},\varepsilon_\lambda\middle|\tilde{G}_0\middle|N_f;\bar{k},\varepsilon_\lambda\right\rangle = \left[z - E_N - \omega - C(z-\omega)\right]^{-1}$ are the diagonal elements of the free Green operators before and after the transition in the Furry picture and E_N is the energy of the giant nucleus ($E_N = 2E_U + \Delta M$, ΔM is the mass excess). $C(z)$ is the self-energy function of the giant double-uranium nucleus. Substituting (12) into (11), we obtain the following form of the evolution operator:

$$\left\langle N_f;\bar{k},\varepsilon_\lambda\middle|U(t,0)\middle|N_i\right\rangle = \frac{1}{2\pi}\int\frac{\left\langle N_f;\bar{k},\varepsilon_\lambda\middle|M(z)\middle|N_i\right\rangle}{\left(z - 2E_u - \omega\right)\left(z - E_N - \omega - C(z-\omega)\right)\left(z - E_N - C(z)\right)}e^{-izt}dz. \quad (13)$$

Here we have taken into account that the most contribution to the self-energy function of the supercritical atom comes from processes of the interaction of its nucleons with the vacuum and as a consequence it is approximately equal to the self-energy function $C(z)$ of the nucleus. At the first stage we can take $M(z)$ to be equal to the electromagnetic interaction Hamiltonian $M(z) = H_I = e\int j_\mu A^\mu d^3x$, which is of the following form for dipole transitions: $\langle N_f; \bar{k}, \varepsilon_\lambda | H_I | N_i \rangle = \sqrt{2\pi\omega_{fi}} \langle N_f; \bar{k}, \varepsilon_\lambda | \bar{n} \times \hat{d} | N_i \rangle$, $\hat{d}$ being the dipole moment operator, $\bar{n} = \bar{k} / \omega_{fi}$ and ω_{fi} is the atomic transition frequency. Integrating (11) over z and Ω, summing over λ, we obtain the following expression:

$$P(\omega) = \frac{A}{|\Delta M + C(2E_U)|^2} \frac{\omega}{|(\omega - \Delta M - C(2E_U + \omega))|^2}, \tag{14}$$

where A is a new normalization factor: $A = 4\pi A' |V_{fi}|^2$, with $|V_{fi}|^2 = 2\pi\omega_{fi} |\langle N_f; \bar{k}, \varepsilon_\lambda | \hat{d} | N_i \rangle|^2$. The behavior of the self-energy function $C(z)$ near the point $z = E_N$ determines the decay law of the unstable nucleus. In general this law is not exponential one and is characterized not only by a decay width but also by other parameters determining the behavior of $C(z)$ near the point $z = E_N$:

$$\mathrm{Re}\, C(z) = (z - E_n) B_1, \ \mathrm{Im}\, C(z) = \Gamma_0 + (z - E_n) B_2, \tag{15}$$

where Γ_0 is the decay width of the giant double-uranium nucleus, $B_1 = d\,\mathrm{Re}\, C(z) / dz|_{z=E_n}$ and $B_2 = d\,\mathrm{Im}\, C(z) / dz|_{z=E_n}$. Here we have taken into account that $\mathrm{Re}\, C(z = E_n) = 0$ because of the fact that we deal with the renormalized nucleus mass. The dimensional analysis of the GDE applied to the problem shows that $B_1 = \frac{\mu}{\Delta M} \cdot I_1$ and $B_2 = \frac{\mu}{\Delta M} \cdot I_2$, where I_2 is of order $O(1)$. The dimensionless parameter I_1 should be much smaller and will be neglected in further discussion. Taking into account that ΔM is much smaller than the reduced mass μ of two Uranium nuclei, we finally get

$$C(2E_U + \omega) = i\Gamma_0 + i\frac{\mu}{\Delta M} \cdot I \cdot \omega \tag{17}$$

with $I \equiv I_2$. Substituting expression (17) into (14) we obtain:

$$P(\omega) = \frac{A}{\Delta M^2 + \Gamma_0^2} \frac{\omega}{(\omega - \Delta M)^2 + \left(\Gamma_0 + \frac{\mu}{\Delta M} \cdot I \cdot \omega\right)^2}. \tag{18}$$

The slope of curve (18) is as follows:

$$\frac{dP(\omega)}{d\omega} = \frac{A}{\Delta M^2 + \Gamma_0^2} \left\{ -\omega^2 \left[1 + \left(\frac{\mu}{\Delta M} \cdot I \right)^2 \right] + \Delta M^2 + \Gamma_0^2 \right\} \left[(\omega - \Delta M)^2 + \left(\Gamma_0 + \frac{\mu}{\Delta M} \cdot I \cdot \omega \right)^2 \right]^{-2} \tag{19}$$

Taking into account that for the double uranium nucleus the theoretically predicted mass excess is $\Delta M \approx 9 \cdot 10^8 eV$ [3], $\mu_{^{238}U} \approx 1.1 \cdot 10^{11} eV$, $\frac{\mu}{\Delta M} \cdot I \approx 1.2 \cdot 10^2$, and the optical photon energy is $\omega \sim 2\mathrm{eV}$, we can neglect 1 in comparison with $\frac{\mu}{\Delta M}\cdot I$ in the numerator and ω in comparison with ΔM in the denominator. Equation (19) then reads:

$$\frac{dP(\omega)}{d\omega}=\frac{A}{\Delta M^2+\Gamma_0^2}\frac{-\omega^2\left(\frac{\mu}{\Delta M}\cdot I\right)^2+\Delta M^2+\Gamma_0^2}{\left[\Delta M^2+\left(\Gamma_0+\frac{\mu}{\Delta M}\cdot I\cdot\omega\right)^2\right]^2}. \quad (20)$$

The greatest quantity here is the mass excess ΔM. The spectrum will be determined substantially by this parameter and hence may serve an experimental test for its theoretical value. The contribution from the nuclear decay width Γ_0 is smaller, but still very important. Let us consider two cases. If the decay time is $\tau_N \sim 10^{-21}$ s, as it is predicted by nuclear physics, then the decay width is $\Gamma_0 \approx 5\cdot 10^5$ eV, and its contribution to the spectrum will be essential. In particular, we can neglect the term $\frac{\mu}{\Delta M}\cdot I\cdot\omega$ in comparison with the decay width in (20). Otherwise, if $\tau_N \sim 10^{-19}$ s, as it is desirable for the spontaneous vacuum decay to occur, then the decay width is $\Gamma_0 \approx 5\cdot 10^3$ eV. In this case Γ_0 and $\frac{\mu}{\Delta M}\cdot I\cdot\omega$ are comparable, and initial expression (20) should be used.

5. Conclusion

We have obtained the shape of a spectrum line emitted by an atom with a superheavy nucleus. It is mainly determined by the mass excess ΔM of the giant nucleus, and hence is an important tool for experimental test of the theoretical value of this parameter. The nuclear decay parameters Γ_0 and I are determined by the form of the light part spectrum that looks as continuous frequency spectrum. Thus, optical methods provide an additional possibility of independent estimation of the superheavy nucleus characteristics, such as the mass excess and the decay parameters, the ratio of which is crucial for the experimental observation of the spontaneous vacuum decay. Although the energy of the photons with which we deal in the case is very small, here it is multiplied by the factor 100, and hence the magnitude of the response will be large enough to be observable on the background. Moreover the differential spectrum is considered, since it is known to provide greater accuracy. The method presented in this paper is universal and may be applied for the study of other unstable nuclei.

References

[1] Hänsch T W 2006 *Rev. Mod. Phys.* **78** 1297
[2] Antognini A *et al* 2013 *Science* **339** 417
[3] Greiner W, Müller B and. Rafelski J 1985 *Quantum Electrodynamics of strong fields* (Berlin: Springer)
[4] Leinberger U *et al* 1997 *Phys. Lett. B* **394** 16
[5] Ahmad I *et al* 1999 *Phys. Rev. C* **60** 064601
[6] Heinz S 2005 *Topics in Heavy Ion Physics Memorial Symposium for Gerhard Soff (Frankfurt am Main)* (Budapest: EP Systema) p 57
[7] Gainutdinov R Kh 1999 *J. Phys. A: Math. Gen.* **32** 5657
[8] Gainutdinov R Kh, Mutygullina A A and Scheid W 2002 *Phys. Lett. A* **306** 1
[9] Gainutdinov R Kh and Mutygullina A A 2002 *Phys. Rev. C* **66** 014006
[10] Gainutdinov R Kh, Moutygoullina A A 1997 *Physics of Atomic Nuclei* **60** (5) 841
[11] Gainutdinov R Kh, Moutygoullina A A 1999 *Physics of Atomic Nuclei* **62** (11) 1905

[12] Gainutdinov R Kh, Moutygoullina A A 2002 *Physics of Atomic Nuclei* **65** (8) 1421
[13] Gainutdinov R Kh and Mutygullina A A 2009 *Physica Scripta* T **135** 014012
[14] Gainutdinov R K 1989 *J. Phys. A: Math. Gen.* **22** (3) 269
[15] Gainutdinov R Kh 1995 *Zh. Èksp. Teor. Fiz.* **108** 1600

Electron dynamics at GaAs-AlGaAs heterojunction studied by ultrafast spectroscopy

A V Leontyev[1], K V Ivanin[1], T G Mitrofanova[1], V S Lobkov[1] and V V Samartsev[1,2]

[1]Zavoisky Physical-Technical Institute, 10/7 Sibirsky tract 420029 Kazan Russia
[2]Kazan Federal University, 18 Kremlyovskaya st. 420008 Kazan Russia

E-mail: mailscrew@gmail.com

Abstract. In this letter the electron and spin dynamics at GaAs/AlGaAs heterojunction was studied by ultrafast spectroscopy techniques (photon echo and transient grating studies). Relaxation times and diffusion coefficients of photoexcited electrons and spins were obtained using pure optical setup. The estimated spin diffusion coefficient value of 160 cm^2/s is relatively high and comparable to the electron diffusion coefficient of 200 cm^2/s. This feature makes GaAs/AlGaAs heterosructure a promising material for practical application in semiconductor spintronics.

1. Introduction

The coherent excitation of a semiconductor structure leads to a nonlinear polarization which later decays on a subpicosecond timescale because of phase-breaking scattering processes affecting photoexcited carriers and phonons. A possible way to study their interactions involve the transient coherent spectroscopy using ultrashort laser pulses [1],[2]. While the optical dephasing and relaxation experiments can be used to investigate the fundamental static and dynamic properties of condensed matter, they also favour the progress in electronic charge- or spin-based components [3],[4].

The semiconductor heterostructure we studied consists of several layers epitaxially grown on GaAs substrate (figure 1). The heterojunctions of this kind show remarkably high carrier mobilities due to the separation of the free carriers from the parent ionized donors. A large amount of free carriers accumulate at the interface forming a highly populated two-dimensional 10 nm thick electron layer.

2. Transient electron and spin gratings in GaAs / AlGaAs heterostructure

When the two identical pump pulses of wavelength λ simultaneously cross at an angle θ in the sample their interference produces a standing wave in the plane of the sample. In case both the pump beams are polarized linearly and parallel to each other, they interfere and generate a spatially periodic electron concentration grating. Note that the internal electric field near the heterojunction area shoves the holes into the bulk GaAs, and thus the observed transient grating is electron-only.

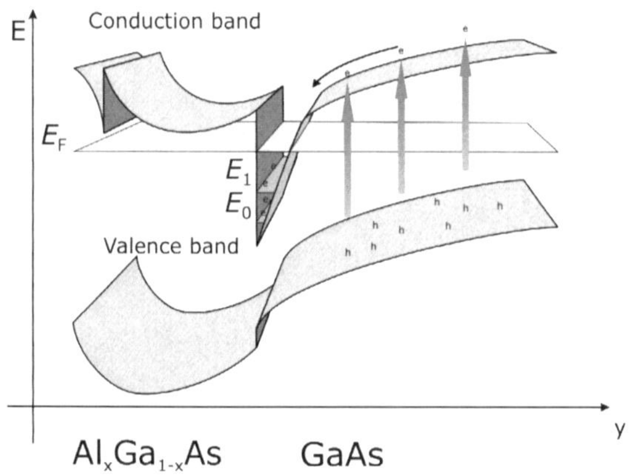

Figure 1. Energy diagram of GaAs/AlGaAs heterostructure.

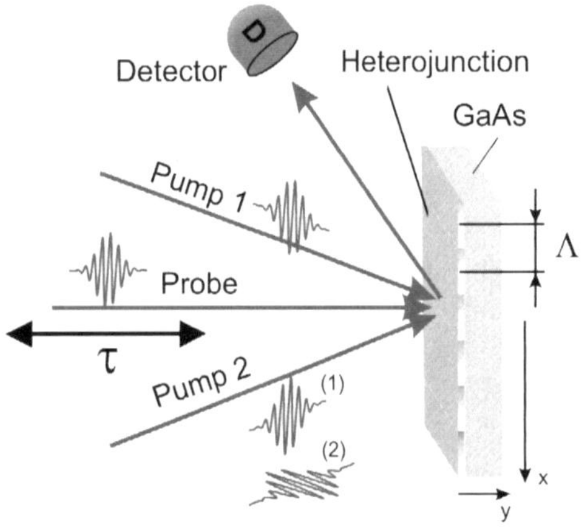

Figure 2. Diffraction of the probe pulse on a population(1) or spin(2) grating.

In the simplest model, assuming the main role of relaxation and diffusion in the transient grating destruction, the process is described by [5]

$$\frac{\partial \Delta N(x,t)}{\partial t} = -D\nabla(\Delta N(x,t)) - \frac{\Delta N(x,t)}{T_r}, \tag{1}$$

where T_r is the relaxation lifetime, D is the coefficient of the electron diffusion. In (1) only a linear relaxation lifetime is included and the diffusion coefficient is assumed to be density independent.

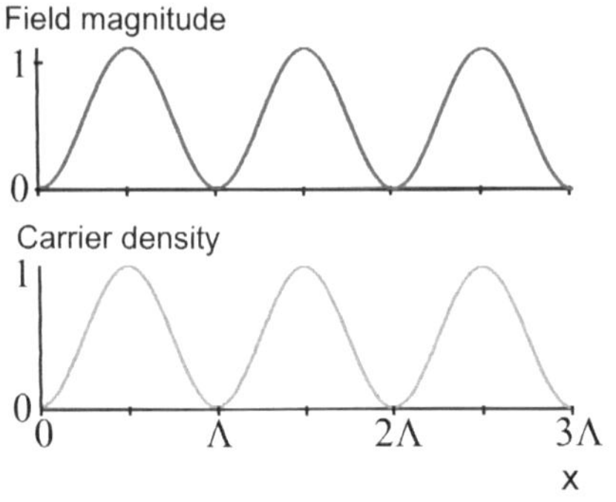

Figure 3. Co-polarized pump beams: field magnitude distribution and photoexcited electron density at the sample plane.

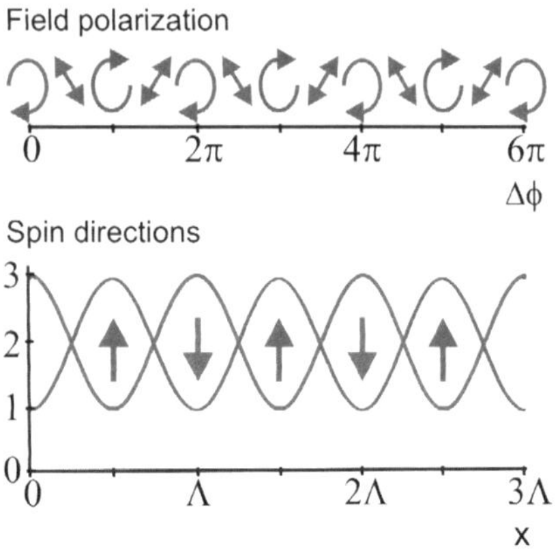

Figure 4. Cross-polarized pump beams: polarization distribution and spin directions at the sample plane.

The grating period can be calculated as $\Lambda = \lambda/(2\sin\theta/2)$, then the electron grating decay rate is expressed in terms of Λ by

$$\frac{1}{T_{gr}} = \frac{8\pi^2}{\Lambda^2}D + \frac{2}{T_r}. \tag{2}$$

If the pump beams are cross-polarized the electromagnetic field is spatially distributed like

$$I = I_+ \cos^2(\phi + \pi/2) + I_- \sin^2(\phi + \pi/2) \tag{3}$$

or the superposition of two left and right circularly polarized standing waves shifted with respect to each other by $\Lambda/2$, and the spin direction distribution would follow this pattern. So instead of sinusoidal transient grating of electron population we produce two spin orientation gratings, while the carrier distribution would be uniform.

If the diffusion-driven grating decay plays the substantial role, then the decrease of Λ would shorten the grating lifetime. The latter can be extracted through measurement of the diffracted light from a delayed probe pulse. In figure 5 the grating decay rate is plotted versus $8\pi^2/\Lambda^2$. According to (2), the slope of a line approximation is the diffusion coefficient and the intercept is proportional to the electron relaxation rate $1/T_r$.

This fitting procedure yields an electron diffusion coefficient of 200 cm^2/s and spin diffusion coefficient of 160 cm^2/s at 295 K. The relaxation times were estimated as 3 ns and 50 ps for electrons and spins, respectively.

3. Coherent four-wave mixing spectroscopy of GaAs / AlGaAs heterostructure

Unlike disperse atomic systems in which the optical response can be determined by independent transitions between atomic levels, in semiconductor heterostructures saturation effects of correlations and nonlinear dynamics due to the Pauli principle are most significant.

For the transverse relaxation time evaluation, we follow the degenerate four-wave mixing technique, introduced by Weiner and Ippen [6]. In this setup the delay $\Delta\tau$ between the pump pulses is being altered. The echo signal is emitted and measured in momentum-matched direction.

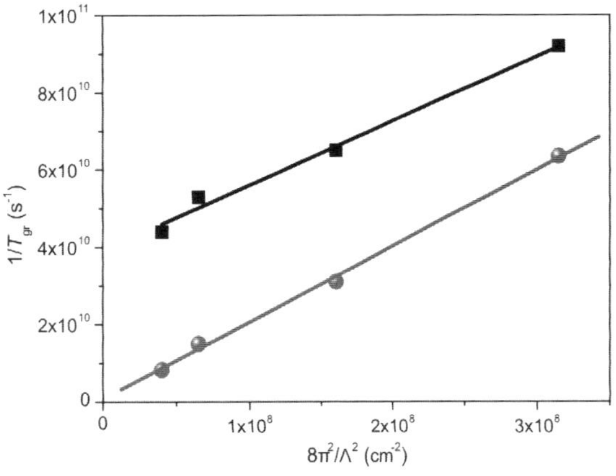

Figure 5. Population (●) and spin (■) grating decay rate versus $8\pi^2/\Lambda^2$

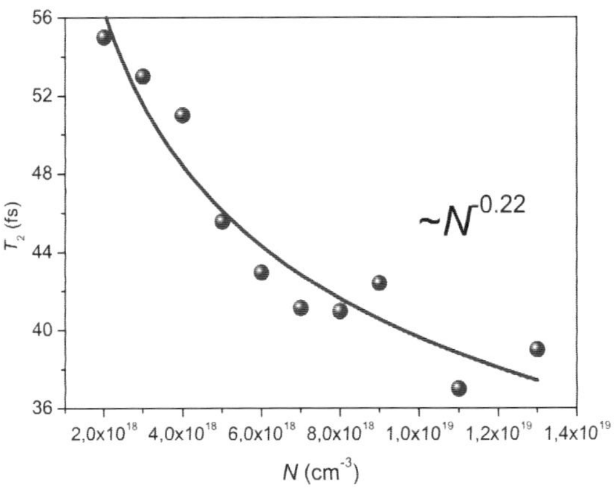

Figure 6. The dephasing time T_2 as a function of carrier density N.

According to [7], the recorded signal varies with $\Delta\tau$ like

$$I(\tau) \propto \int_0^\infty \mathrm{d}t \int_0^\infty \mathrm{d}t' f(t'-t) f(t-\Delta\tau) f(t'-\Delta\tau) exp[-2(t+t')/T_2], \tag{4}$$

$f(t)$ is the autocorrelation function of laser pulses.

The four-wave mixing measurements and approximation the data by (4) showed the T_2 change from 55 to 40 fs caused by the carrier density increase. The power-law fitting of this dependence yielded the exponent of -0.22 ± 0.02 which evidences the presence of Coulomb electron-electron interaction screening. This value differs from -0.3 obtained for bulk GaAs as reported in [8, 9]. One of the crucial points of optoelectronics and optical computing at the present stage of studies is the research of the promising materials possessing the coherent state lifetime as long as possible. An important parameter from the practical point of view is the sample temperature. The room temperature could be the best option as it eliminates the need of the sample cooling. The comparison of the experiment results on bulk GaAs [8] and heterostructure sample in this letter shows that the dephasing time in heterostructure is approximately 3 times higher than that in bulk GaAs in the similar conditions (e.g. 55 fs versus 20 fs at the carrier density of 2×10^{18} cm^{-3} and the room temperature).

4. Conclusion

We have presented a four-wave mixing and transient grating experiment on the GaAs/AlGaAs heterojunctionon a subpicosecond and picosecond timescale. The electron and spin diffusion coefficients of 200 cm^2/s and 160 cm^2/s, respectively, were obtained. Since the spin diffusion coefficient is comparable to the carrier diffusion coefficient, the structures of this type are suitable for spin transport applications at room temperature.

Acknowledgments

This study was supported by the Program of RAS Presidium "Quantum mesoscopic and disordered structures", the Programs of the Department of Physical Sciences of RAS "Fundamental spectroscopy and its applications", the Program "Leading Scientific Schools" (No NSh.5602.2012.2), and by RFBR grants (No 12-02-31381-mol_a, No 14-02-00041-a, No 14-03-00331-a, No 14-03-31287-mol_a and No 14-02-90000_Bel-a).

References

[1] Weiner A 2011 *Ultrafast Optics* (New York: Wiley)
[2] Ivanin K V, Leontiev A V, Lobkov V S, Nikiforov V G, Salikhov K M, Samartsev V V and Safiullin G M 2009 *Laser Phys. Lett.* **6** 644
[3] Wolf S, Awshalom D D, Buhrman R A, Daughton J M, von Molnár S, Roukes M L, Chtchelkanova A Y and Treger D M 2001 *Science* **1** 1488
[4] Zutic I, Fabian J and Das Sarma S 2004 *Rev. Mod. Phys.* **76** 323
[5] Moss S C, Lindle J R, Mackey H J and Smirl A L 1981 *Appl. Phys. Lett.* **39** 227
[6] Weiner A M and Ippen E P 1984 *Opt. Lett.* **9** 53
[7] Othonos A 1998 *Appl. Phys. Lett.* **84** 1789
[8] Becker P C, Fragnito H L, Brito C H, Fork R L, Cunningham J E, Henry J E and Shank C V 1988 *Phys. Rev. Lett.* **61** 1647
[9] Wegener M and Hugel W 2001 *Adv. in Solid State Phys.* **41** 89

Implementation of double-pulse laser control in optical Kerr effect spectroscopy

V G Nikiforov[1], A G Shmelev[1], G M Safiullin[1], V S Lobkov[1] and V V Samartsev[1,2]

[1]Zavoisky Physical-Technical Institute, Sibirsky tract 10/7, 420029, Kazan, Russia
[2]Kazan Federal University, Kremlyovskaya St. 18, 420008, Kazan, Russia

E-mail: sgartjom@gmail.com

Abstract. Two-pulse control of time-dependent anisotropy in liquid CCl_4 and $CHCl_3$ at room temperature is implemented using femtosecond polarisation spectroscopy. Non-resonant excitation was enhanced by means of the double-pulse pump-probe technique. It is shown that by varying the delay between the exciting pulses and their relative intensity, selection of contributions of individual intramolecular modes into the recorded signal is achieved. The molecular responses were detected using the time-resolved optically heterodyne-detected optical-Kerr-effect technique.

1. Introduction

Laser control of the medium's state is one of the research priority areas focused on theoretical and experimental investigations [1]. Currently, the most effective instrument for the implementation of laser control is femtosecond technique, which makes it possible to excite a system by pulses comparable (in duration) to the period of molecular oscillations [2, 4, 3]. One of the laser control schemes is based on the use of multipulse excitation. The wave packets in the medium are formed by timed sequences of femtosecond pulses [5]. The key to excite a certain state of the medium is the optimized parameters of the multiple-pulse excitation (the pulse intensities, the delays between the pulses, etc.).

In this study we use the double-pulse non-resonant excitation to control the amplitudes of the individual modes in the vibrational responses of the molecules in liquid. We suppose that because of the short relaxation time of the coherent molecular motions (of less than or a few picoseconds), this technique was not employed in liquid at room temperature. In general, the pump pulse provides the vibrational and rotational wave packets due to which the anisotropy of the polarizability is induced in liquid. The evolution of the polarizability is probed by the time-resolved optically heterodyne-detected optical Kerr effect (OHD-OKE) [2, 3, 4]. The femtosecond pulses were generated by a standard self-mode-locked Ti:Sa laser (Avesta-Project, Troitsk, Russia). The average energy of the laser pulses on the carrying wavelength of 790 nm was 4×10^{-9} J. This pulses can't excite modes higher than 500 cm^{-1}. In our experiment the two pump pulses are independently adjustable, and a standard scheme of the OHD-OKE transient detection [2, 3] was used: the probe beam polarization was at 45° with respect to the pump beam polarization, the intensity of the probe beam passed through the polarizer, the sample

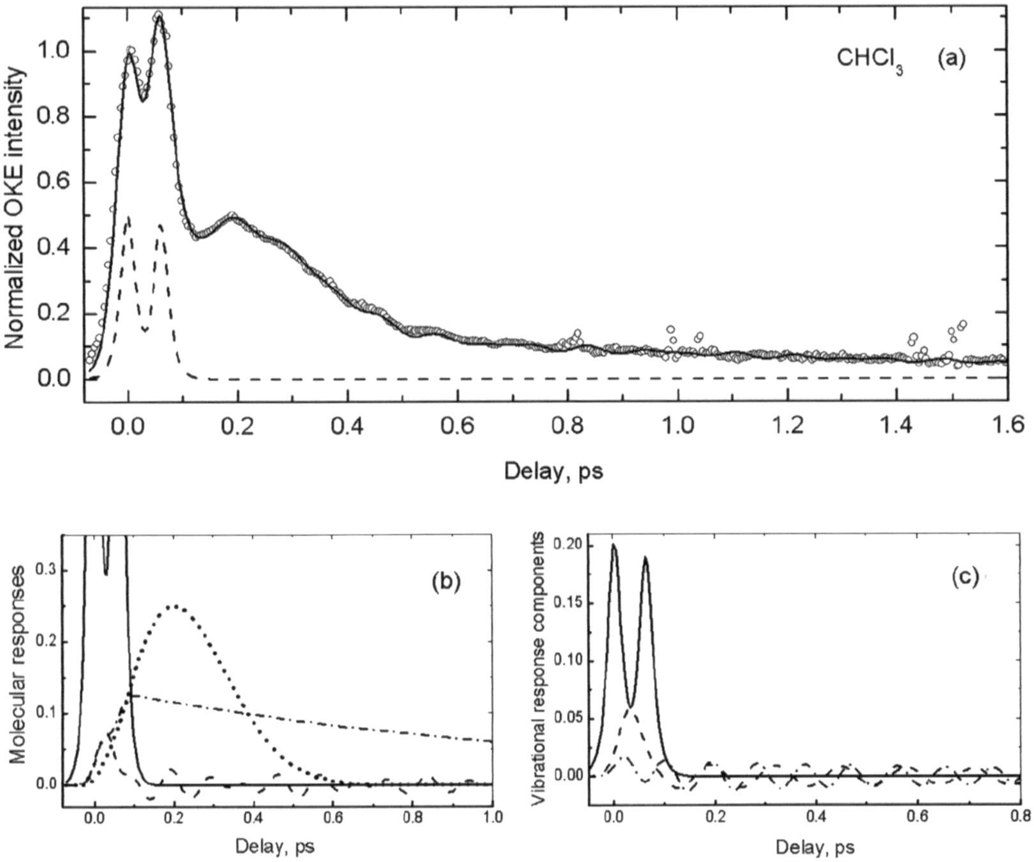

Figure 1. The double-pulse excitation of the OHD-OKE signal. Delay between the pump pulses is 60 fs and their intensity ratios 0.94. (a) O – the experimental data, solid line – the simulation, dashed line – the envelope of the double-pulse intensities. (b) Solid line – the electronic response, dashed line – the vibrational response, dotted line – the librational response, dash-dotted line – the orientational response. (c) Solid line – the envelope of the double-pulse intensities, dashed line – the vibrational response component of the 258 cm^{-1} mode, dash-dotted line – the vibrational response component of the 362 cm^{-1} mode.

and the analyzer was measured. The optical heterodyne detection and the lock-in amplifier were used. We also used simulation of motion responses described in [5, 6].

2. Results

Chloroform has two Raman-active vibrational modes at frequences less than 400 cm^{-1}: 258 and 362 cm^{-1}. One can see suppression of 362 cm^{-1} mode by second pulse in figure 1. Registered signals are superposition of responses of individual motions [3] : instantenious response of electrons and vibrational, librational and orientational response of nuclear skeleton of molecule.

By varying time delay between pump pulses and their relative intensity we can suppress or enhance responses of individual motions. CCl_4 has three low-frequency Raman-active vibrational modes: 458 cm^{-1}, 312 cm^{-1} and 216 cm^{-1}. One can see suppresion of responses of individual vibrational modes in figure 2, figure 3 and figure 4. In the figure 2 one can see that responces of individual vibrational modes after second pulse excitation changes in 1.8

(216 cm^{-1}), 0.2 (312 cm^{-1}) and 0.5 (458 cm^{-1}) times. With the same relative intensities of pump pulses, but with time delay 845 fs between pulses after second excitation responses changes in 0.2 (458 cm^{-1}), 1.9 (312 cm^{-1}) and 0.5 (216 cm^{-1}) times (figure 3). With the time delay of 1025 fs responses changes in 0.6 (216 cm^{-1}), 0.2 (312 cm^{-1}) and 2 (458 cm^{-1}) times (figure 4).

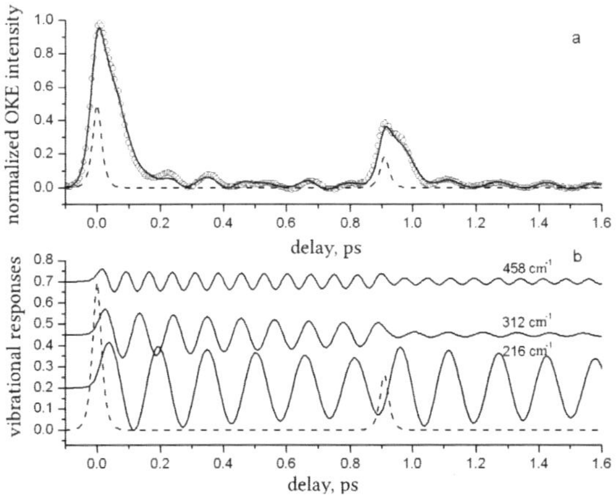

Figure 2. OKE signal in liquid CCl_4 under double-pulse excitation, delay between pump pulses 908 fs, intensity ratio 0.38): (a) ○ – the experimental data, solid curve – simulation, dashed curve – envelope of the intensity of the exciting pulse. (b) simulation results of the vibrational responses, dashed curve is the envelope of the intensity of the exciting pulse.

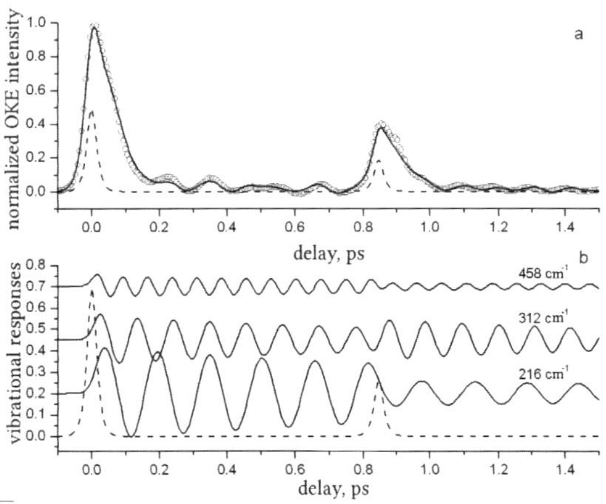

Figure 3. OKE signal in liquid CCl_4 under double-pulse excitation, delay between pump pulses 845 fs, intensity ratio 0.38): (a) ○ – the experimental data, solid curve – simulation, dashed curve – envelope of the intensity of the exciting pulse (b) simulation results of the vibrational responses dashed curve – envelope of the intensity of the exciting pulse.

3. Conclusion

These experiments present an example of the double-pulse controlling of the coherent molecular vibrational dynamics in liquid at room temperature. The laser control makes it possible to manipulate the vibrational responses in the OHD-OKE signal. The key parameters of the double-pulse scenarios increase or decrease the amplitude of the vibrational response and separate the desired modes in vibrational response are time delay between pump pulses and their relative intensity. Vibrational response of the mode 458 cm^{-1} is not registered, because probe pulse width is $\propto$ 400 cm^{-1}. However, simulation indicates that the double-pulse excitation also controls the intensity of the response of the given mode. Another important fact is suppression of the vibrational mode response at delays comparable to the relaxation time requires a decrease in the intensity of the second pump pulse. Note also that in liquids, where intensive rotational-vibrational responses are registered along with vibrational responses in the OKE signal, the scenario of complete suppression of the vibrational response at the lowest possible delay between

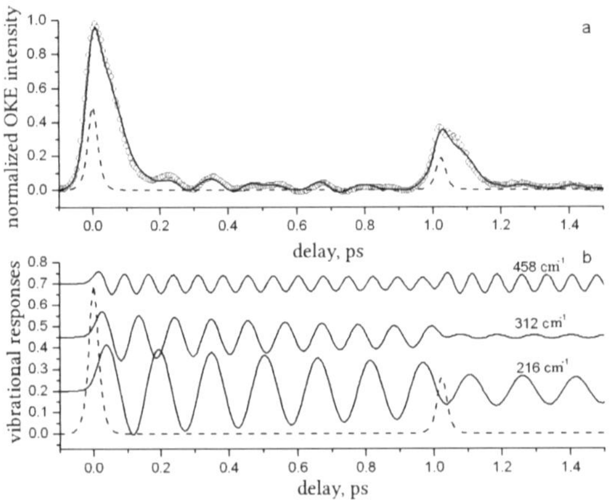

Figure 4. OKE signal in liquid CCl_4 under double-pulse excitation, delay between pump pulses 1025 fs, intensity ratio 0.38): (a) O – the experimental data, solid curve – simulation, dashed curve – envelope of the intensity of the exciting pulse (b) simulation results of the vibrational responses dashed curve – envelope of the intensity of the exciting pulse.

the exciting pulses can be used to extract the rotational response. The contributions due to the molecular rotations increase as a result of the impact of the second pump pulse at any delay.

Acknowledgments
This study was supported by the Programs of the Department of Physical Sciences of RAS "Fundamental spectroscopy and its applications", the Program "Leading Scientific Schools" (No NSh.5602.2012.2), and by Russian Foundation of Basic Research (projects No 12-07-97017-r_povolzh'e, No 12-03-31313-mol_a, and No 14-02-90000-Bel_a).

References

[1] Steinmeyer G 2003 *J. Opt. A. : Pure Appl. Opt.* **5** R1
[2] Ruhman R, Williams L R, Joly A G, Kohler B and Nelson K A 1987, *J. Phys. Chem.* **91** 2237
[3] McMorrow D, Lotshaw W T and Kenney-Wallace G A 1988 *IEEE J. Quantum Electron.* **24** 443
[4] Righin R 1993 *Science* **262** 1389
[5] Moiseev S A and Nikiforov V G 2004 *Quantum Electron* **34** 1077
[6] Nikiforov V G and Lobkov V S 2006 *Quantum Electron* **36** 984

Heat-assisted effects in ferromagnetic nanoparticles

A A Bukharaev[1,2], D A Bizyaev[1], T F Khanipov[1,2], N I Nurgazizov[1] and A P Chuklanov[1]

[1] Zavoisky Physical-Technical Institute, 420029, Russian Academy of Sciences, Kazan, Tatarstan, Russia
[2] Kazan (Volga Region) Federal University, 420008, Kazan, Tatarstan, Russia

E-mail: a_bukharaev@kfti.knc.ru

Abstract. The processes of the magnetization reversal by the external magnetic field of the Py particles have been studied by magnetic force microscopy in the temperature range of 300-650 K. The values of the switching field of the particle magnetization and the switching field distribution for the particles array have been determined. The switching field and the field distribution decrease significantly with increasing temperature.

1. Introduction

The analysis of modern technologies and experimental results makes it possible to conclude that one of the most perspective methods of data recording is heat-assisted magnetic recording (HAMR) [1-3]. In HAMR, data recording is carried out by the magnetic switching of the local area by the magnetic field of a recording head with additional heating of the area by the focused laser beam. The data density can be increased by using a medium composed of separately located ordered one-domain ferromagnetic nanoparticles (so-called patterned medium [3]). The theoretical limit of the data density achieved by HAMR is 10 Tb/in^2. Particles with two opposite orientations of the magnetization state corresponding to "0" and "1" are used for data recording in a binary code. Switching of the magnetization state of the one-domain nanoparticle is a physical base of data recording in the same devices. To implement HAMR, it is important to know the value of the switching field (SF) for the particle at the temperatures much higher than room temperature (up to the Curie temperature) because the coercive force of a ferromagnetic material is decreased by 3-4 times at heating [2, 4]. The control of the switching field distribution (SFD) of different particles is also essential for HAMR [5].

Magnetic force microscopy (MFM) is a convenient technique for studying the magnetization reversal of a single nanopraticle or particle arrays. Information about switching fields depending on the material, shape and sizes of particles was already obtained by MFM [5, 6]. The temperature dependence of magnetic reversal processes was usually studied in the temperatures range of 10-300 K [7, 8]. In this work, MFM was used for studying the magnetization reversal of the Py particles at temperatures from 300 to 650 K. The change of SF and SFD with increasing temperature was also estimated and the magnetization reversal mechanism was analyzed by using the computer simulation of MFM images.

2. Sample preparation and measurement techniques

The planar structures with the isolated ferromagnetic Py particles on SiO_2 surface was manufactured using the modified technique of scanning probe nanolithography (SPNL) described in [9]. According to this technique, the lithographic mask was formed in a layer of poly-(methylmethacrylate) (PMMA) located on the SiO_2 surface by the probe of an atomic force microscope (AFM) Solver P47 (NT-MDT). At the next step, a ferromagnetic metal film was sputtered on the mask and the sacrificial layer was removed by the lift-off technique. The cantilevers NSG20 (NT-MDT, force constant 48 N/m) and NSC15 (MikroMasch, force constant 46 N/m) with the tip curvature radii of 10 nm were used. The PMMA film (50-60 nm) spin-coated on the substrate was used. The sample with the lithographic mask was coated by Py (Ni 78%, Fe 22%) at ultra high vacuum conditions (10^{-8} mbar) using an electron beam evaporator built into the UHV system Omicron Multiprobe P. The deposition rate was about 0.1 nm/min. Then the sample was rinsed in an ultrasonic bath with chlorbenzene to allow the lift-off of the sacrificial layer.

The AFM image of 30 submicron Py particles on the SiO_2 surface is presented in figure 1a. The lateral size of particles was in the range from 400x150 to 600x250 nm, the height of particles was 15 nm. The MFM image of the same area is shown in figure 1b. The SF dependence due to the range of the lateral size of particles and the SFD dependence on increase in the temperature were studied. The Solver HV system (NT-MDT) was used to study the magnetization reversal of the Py particles. The Solver HV system makes it possible to obtain AFM and MFM images and to heat a sample in the vacuum conditions and to apply an external magnetic field in the sample plane direction. The NSC19/Co-Cr (MikroMasch) magnetic tips were used and the phase change of oscillated cantilever was measured. The one-pass technique of MFM image registration was used. In one pass technique the magnetic probe scans a sample on the advance set height necessary to no magnetizing a particle by a probe field.

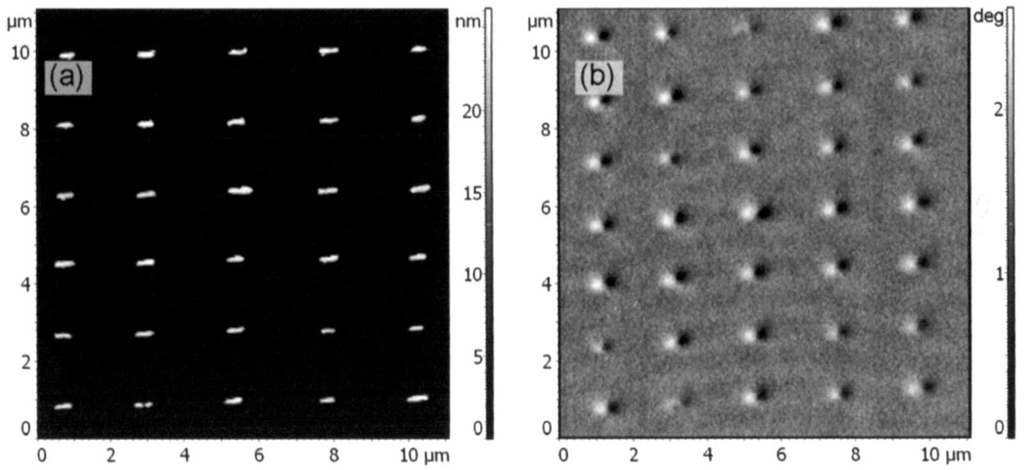

Figure 1. AFM image of the array of 30 Py particles (a), MFM image of the same area obtained by the one-pass technique after particles were magnetized in one direction by the external magnetic field (b).

The magnetization reversal of Py particles at room temperature was studied in the same manner as in [10, 11]. At the first stage all particles were magnetized in the sample plane by the external field (of about 800 Oe) appreciably higher than the coercive force of Py particles. After that the magnetic states of all particles was checked by MFM (figure 1b). Then the magnetic field of 100 Oe was applied to the sample in the opposite direction ("magnetization field"). After that the magnetic field was switched off and the magnetization of the same particles by MFM was recorded and the quantity of the particles which have switched magnetization was calculated. The above procedure was repeated with the higher value of the magnetization field. The field was increased step-by-step until all Py particles were switched to the opposite direction. The value of the external field sufficient for the change of the magnetization of the particle to the opposite direction (according to the MFM image of the particle) was fixed as the particle SF. To obtaining the MFM images of the same particles, special markers were created on a sample by the SPNL technique. The cantilever position relative to the markers was observed on a optical microscope.

The remanent magnetization of a separate particle (M) is estimated by the maximum change of the phase of the cantilever oscillation during scan of the particle along the long axis. Our earlier computer simulation of the MFM images [12] showed that such method of the analysis of the MFM images makes it possible to observe the relative change of the magnetization value of a single one-domain particle and to plot the hysteresis loop of the particle.

The particle magnetization at high temperature was studied on the modified Solver HV system. The special (home-built) sample holder with heating up to 800 K was used. To prevent the sample oxidation, the particle magnetizations at high temperature were carried out *ex situ* at the vacuum of about 10^{-2} mbar. At first, the particles were magnetized up to saturation (M_S) by the external magnetic field at room temperature. Then the sample was heated up to the preset temperature and the magnetic field was switched on in the opposite direction. The field magnitude was set significantly smaller than the switching field of particles at room temperature. The heating element was switched off, and the sample was cooled down to the room temperature in the external magnetic field. After that the MFM measurements of the particles array was carried out and the number of the switched particles was calculated. Such technique was used to prevent the MFM probe destruction at the high temperature. At the next step, the experiment was repeated at the higher preset temperature.

3. Results and discussion

According to MFM images, most particles have only two directions of magnetization along the long axis typical for the uniformly magnetized particle (figure 2a). The hysteresis loop of such particles (denoted as "A") obtained by the analysis of the change of MFM images (phase contrast) has a rectangular shape. But some particles during the magnetization reversal process were observed in the intermediate state that differed from the uniform magnetization (figure 2b). In hysteresis loop this state corresponds to the non-uniform magnetization (figure 2b). The step on the hysteresis loop of such particles (denoted as "B") is caused by the formation of a stable state with the low magnetization during the magnetization reversal (figure 2b). Similar results were observed in [6] when the magnetization reversal of the Py particles with the size close to ours and the aspect ratio of 4:1 were studied.

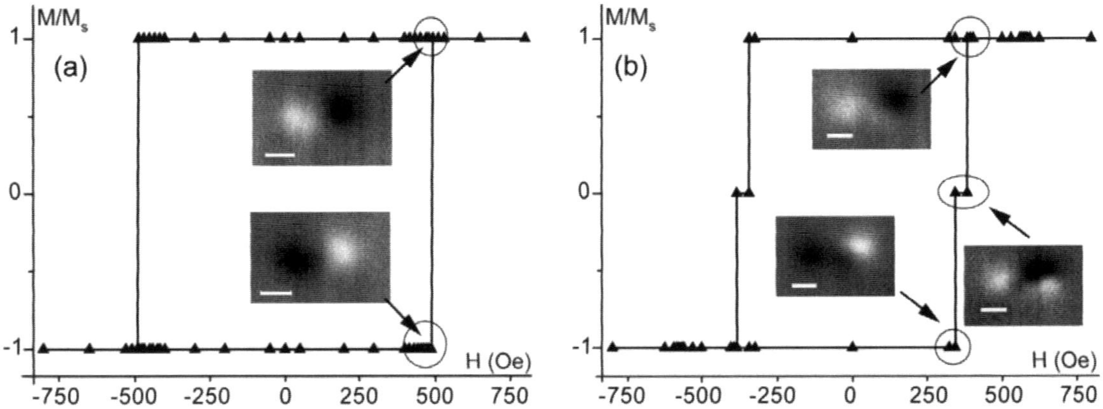

Figure 2. Experimental hysteresis loop for particle "A" (a) and particle "B" (b). The MFM images of the magnetic states insert on curves and the corresponding area indicate by arrows and circles (scale bars, 200 nm).

The computer simulation of magnetization processes of the particles was used to interpret experimental MFM images. The magnetization of particles "A" and "B" was simulated using the

OOMMF [13] program (figure 3a, c). The corresponding MFM images (figure 3b, d) were simulated using the "virtual MFM" program created before [12]. It follows from the analysis of the experimental and simulated MFM images that the non-uniform magnetization is caused by the formation of the vortex magnetization in the middle part of the particle. It is in good agreement with other results of the MFM investigation of the magnetization reversal of Py particles at room temperature [6].

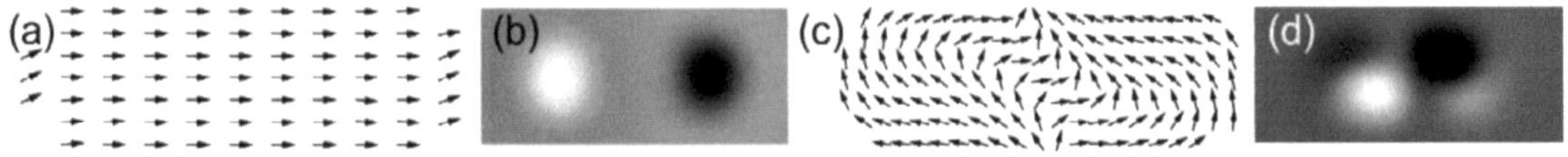

Figure 3. Simulated distributions of the uniform (a) and the non-uniform (c) magnetization state of the Py particles (lateral size of 510x160 nm and 460x190 nm respectively, height of 15 nm) and the simulated MFM images for these particles (b), (d) respectively.

In spite of the small size dispersion, the studied Py particles had significant SF at room temperature (figure 4a). The SF value sufficient for the magnetization reversal of all 30 particles was 680 Oe at room temperature. The increase in the sample temperature led to the significant decreasing of the SF value. The SF value for the same particles was 200 Oe at the temperature of 650 K. The SFD was measured as the difference between the magnetic field required for switching of all 30 particles and the field required for switching of only the first particle. At room temperature the SFD was 220 Oe. The SFD significantly decreased and reached the minimum value of 90 Oe at the temperature increase to 650 K.

Depending on the particle sizes, the magnetization reversal can take place by the coherent rotation of all magnetic moments of particles (Stoner-Wohlfahth model [14]) or by the formation of the intermediate state with the residual vortex magnetization (Neel-Brown model [6, 8]). Since the size of our particles is much larger that the exchange interaction length (of about 10 nm for Py [8]), the coherent rotation of the magnetic moments is improbable. It is obvious that the transition from one state with the uniform magnetization to another state with the opposite magnetization occurs through the formation of the vortex structure (similar to the magnetization reversal in [6]). The same as in work [6], we observed the particles "B" with the stable vortex state. The particles "A" demonstrated only two magnetization states and they had the rectangular shape of the hysteresis loop. It is obviously connected with the stability of the vortex state in a very narrow interval of the external magnetic field. The observation of such state by MFM is improbable because the external magnetic field in the experiments is changed with the limited step (no less than 3 Oe).

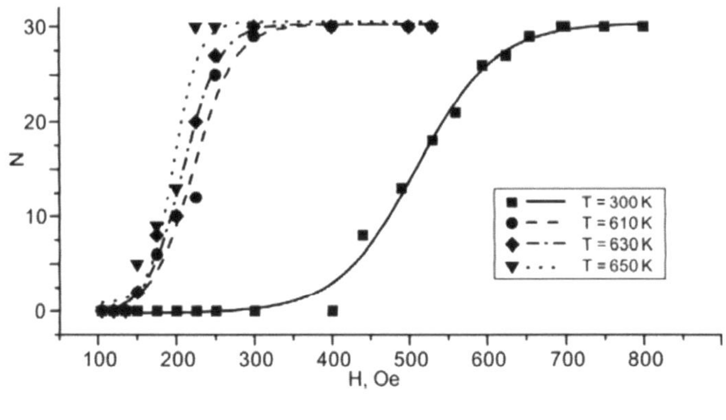

Figure 4. Number of switching particles dependence on the external magnetic field at different temperatures for the array of 30 Py particles.

The decrease in the switching field value with increasing temperature is explained by the known thermo-activation Neel-Brown model of particles switching from one magnetization state to the opposite state [15]. The energy barrier between two stable states of the opposite magnetizations of the particle decreases significantly when the external magnetic field is applied [4, 14], therefore the low field is enough for the magnetic reversal of the Py particle at temperatures of about 500 K. The reduction of the SFD with the increase in the temperature to 500-550 K for particles with the wide size distribution is interesting from the application point of view, because this phenomenon is important for implementing of the high-density data recording using HAMR.

4. Conclusion

The ordered Py particles with the uniform magnetization were manufactured on the SiO_2 surface by the scanning probe nanolithography technique. The processes of the magnetization reversal of the Py particles in the temperature range from 300 to 650 K were investigated by MFM. It was shown the magnetic field necessary for switching from one direction of the uniform magnetization to the opposite direction decreases from 680 to 200 Oe with the increase in the particle temperature to 650 K. The switching field distribution is decreased by more than two times. It follows from the analysis of the experimental and simulated MFM images indicating the structure of the particle magnetization that the switching of the uniformly magnetized particles takes place by the formation of the intermediate vortex structure.

Acknowledgements

This work was supported by a RFBR grant 12-02-00820 and a program of the Russian Academy of Sciences.

References

[1] Inaba Y, Nakata H and Inoue 2011 *Fuji Electric Review* **57** 42
[2] Kryder M, Gage E, McDaniel T, Challener W, Rottmayer R, Ju G, Hsia Y and Erden M 2008 *Proceedings of the IEEE* **96** 1810-35
[3] Şendur K and Challener W 2009 *Appl. Phys. Lett.* **94** 032503
[4] Zeng H, Skomski R, Menon L, Liu Y, Bandyopadhyay S and David J 2002 *Phys. Rev.* B **65** 134426
[5] Tabasum M, Zighem F, De La Torre Medina J, Encinas A, Piraux L and Nysten B 2013 *J. Appl. Phys.* **113** 183908
[6] Zhu X, Grutter P, Metlushko V and Ilic B 2002 *Phys. Rev.* B **66** 024423
[7] Shaw , Russek S, Thomson T, Donahue M, Terris B, Hellwig O, Dobisz E and Shaw M 2006 *Phys. Rev.* B **78** 024414
[8] Shi J, Li J and Tehrani S 2002 *J. Appl. Phys.* **91** 7458
[9] Bizyaev D, Bukharaev A, Lebedev D, Nurgazizov N and Khanipov T 2012 *Technical Physics Letters* **38** 645
[10] Bukharaev A, Ovchinnikov D, Nurgazizov N, Kukovitski E, Klaber M and Wiesendanger R 1998 *Physics Of The Solid State* **40** 1163
[11] Ovchinnikov D, Bukharaev A, Borodin P and Biziaev D 2001 *Physics of Low-Dimensional Structures* **3** 103
[12] Ovchinnikov D and Bukharaev A 2001 *Technical Physics* **46** 1014
[13] http://math.nist.gov/oommf/
[14] Stoner E and Wohlfarth E 1948 *Phil. Trans. R. Soc. Lond.* **A240** 599
[15] Wirth S and von Molnar S 2006 *Handbook of Advanced Magnetic Materials* vol 1 ed Y Liu, D Sellmyer and D Shindo (New York: Springer) pp 294-338

FTIR investigation of molecular mobility in glassy epoxy-amine oligomers with azo-chromophores

K L Nefedyeva[1], A I Fishman[1], M Yu Balakina[2] and T A Vakhonina[2]

[1]Kazan (Volga region) Federal University, 420008 Kazan, Russia
[2]A.E.Arbuzov Institute of Organic and Physical Chemistry KSC RAS, 420088 Kazan, Russia

E-mail: nefedieva_ksu@mail.ru

Abstract. Molecular mobility in glassy epoxy-amine oligomer with covalently attached 4-amino-4'-nitroazobenzene NLO chromophore fragments has been studied by conformational probes method. The conformational transitions in the probes (1,2-dichloroethane and chlorocyclohexane) embedded in the oligomer have been studied in the temperature range from -160°C to +20°C. The freezing of conformational transitions has been observed by FTIR at temperature near T_f = -90°C. This value is correlated with the secondary relaxation transition temperature in the oligomer. The comparison of the activation volumes of the mobile groups in the probes with volumes of mobile groups in the oligomer allows us to conclude that there is the freezing of benzene rings mobility near β-process temperature.

1. Introduction

In the recent decades the interest to organic polymer materials with high quadratic nonlinear-optical (NLO) coefficients raised significantly. This is due to the possibility of their application in photonics and optoelectronics [1]. Molecular sources of NLO response are organic chromophores. Such molecules possess large dipole moments in the ground state. Non-centrosymmetric arrangement of dipole chromophores is formed in an electric field. The second harmonic generation (SHG) is observed in the medium. There are a lot of references concentrated on the investigation of the SHG in polymer matrix with NLO chromophores [1-4]. The great attention is given to the increasing of the relaxation stability of NLO response.

Epoxy polymer matrices based on bisphenol-A diglycidyl ether with aniline fragments in the main chain are promising materials. Organic chromophores are embedded as guest-molecules or covalently attached to the main or side chain of the polymer. The relaxation stability of the NLO response is an important issue related to the mobility of the polymer chain fragments.

In the present work β-process in epoxy-amine oligomer with covalently attached 4-amino-4′-nitroazobenzene NLO chromophore fragments with donor groups introduced into the main chain (**CFAO**) is studied (figure 1).

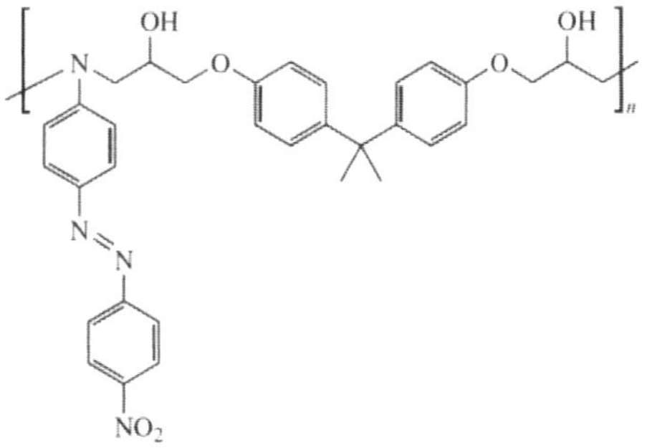

Figure 1. The structure of the monomer unit of CFAO.

2. The method of conformational probes

The molecular mobility of low- and high-molecular-weight glass forming liquids has been extensively studied by different physical methods, including those based on molecular probes. In particular, it was proposed to use compounds with conformational flexibility as probes in glassy solids [5]. The approach is based on embedding a small amount of the probe into the studied glass forming liquid (hereafter "matrix") and following the temperature dependences of its conformationally sensitive bands in FTIR spectra. If at a certain condition the probe molecules can transform from one conformation (A) to another (B) within the observation time, *i.e.*, the conformers are in equilibrium, then the integrated intensities of the IR bands of the conformers, D_A and D_B, follow the relation [6]:

$$\ln\frac{D_A}{D_B}=\frac{\alpha_A}{\alpha_B}+\frac{\Delta S_0}{R}-\frac{\Delta H_0}{RT} \quad (1)$$

where ΔS_0 and ΔH_0 are the entropy and enthalpy differences of the conformers, respectively, α_A and α_B are the absorption coefficients of the conformer's bands under study, T is the temperature and R is the gas constant. The dependence $ln(D_A/D_B)$ upon $1/T$ is shown in figure 2, branch 1.

The conformational transitions in the probe molecules are possible only if the molecules of the surrounding matrix (or their fragments) exhibit sufficient mobility and if such mobility leads to creation of free volume entities large enough for the probe molecules to transform. Lowering the temperature may cause freezing-in of the molecular mobility of the matrix and this in turn may freeze-in the conformational transitions in the probe. Then below a certain temperature T_f the plot of the $ln(D_A/D_B)$ upon $1/T$ should be a straight line parallel to the horizontal axis (figure 2, branch 2). Then, knowing the "activation volume" (*i.e.*, the minimum volume required for conformational transitions to occur), one can estimate the size of the mobile units of the matrix which also froze-in at T_f.

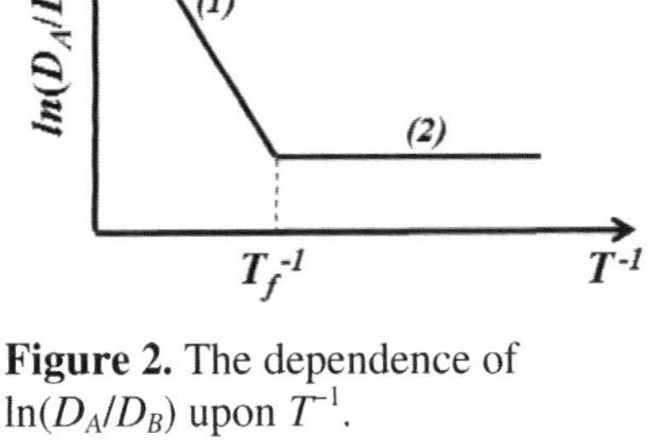

Figure 2. The dependence of $\ln(D_A/D_B)$ upon T^{-1}.

3. Experimental part

The synthesis of **CFAO** is described in [7].

1,2-dichloroethane (**DCE**) and chlorocyclohexane (**CCH**) (products of Sigma–Aldrich) were used as the probes to study the conformational mobility of the oligomer. The polymer films were created from the **CFAO** solution in cyclohexanone. The films were kept in the atmosphere of the **DCE** vapour during few hours. Then the samples were kept in the air during 24 hours. The thickness of the films with DCE was about 30 – 90 μm. The concentration of **DCE**, [C], was estimated to be in the range 1 – 3 vol%. To introduce **CCH** into **CFAO**, mixture of **CFAO** with probe was dissolved in tetrahydrofuran. The liquid solution was spread on KBr plates and kept at room temperature during 24 hours in order to remove the solvent. The concentration of **CCH** in **CFAO** was about 6 vol%, film thickness ca.150 μm.

The absorption bands at 709 cm^{-1} Q(CCl) (*t*-conformation), 653 cm^{-1} Q(CCl) (*g*-conformation) [8] of **DCE** and 684 cm^{-1} Q(CCl) (*a*-conformation), 888 cm^{-1} Q(CC) (*e*-conformation) [9] of **CCH** were chosen as analytical bands. These absorption bands fall within the "transparency windows" of **CFAO**.

The study of conformational dynamics was carried out in the temperature range from -160°C to +20°C. The experimental dependences $ln(D_A/D_B)$ upon $1/T$ for **DCE** dissolved in **CFAO** are represented in figure 3. The enthalpy difference of the conformers was determined for the temperature range -80°C - +20 C. The obtained value appeared to be ΔH_{tg} = 300±40 cal mol^{-1}. The freezing-in of the conformational transitions was observed at T_f = -88±8°C.

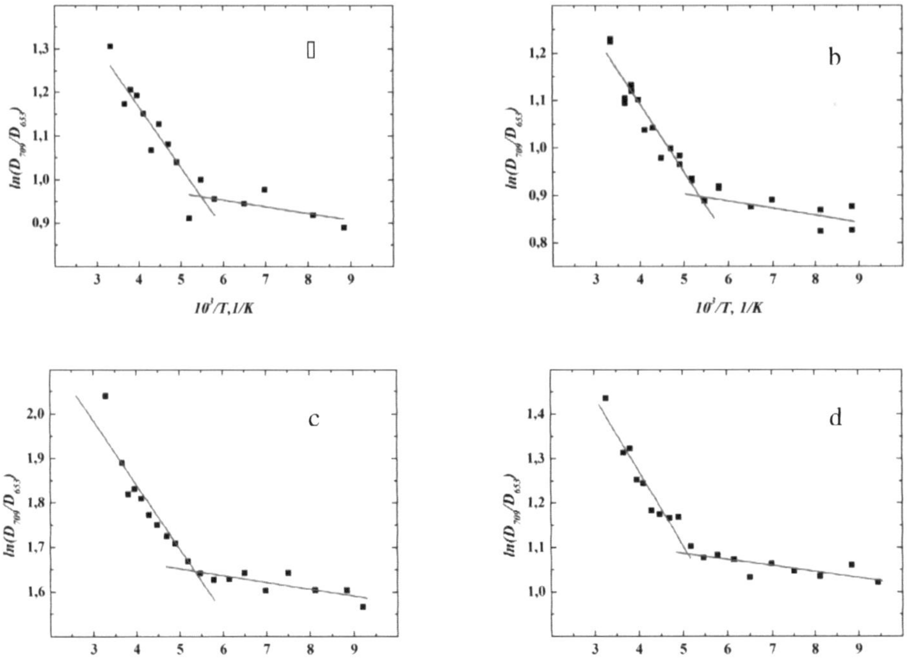

Figure 3. The dependence of $ln(D_t/D_g)$ upon T^{-1} for **DCE** embedded in **CFAO**: a) T_f= -93°C, film thickness d = 90 μm, $[C]$ = 1 vol%; b) T_f= -89°C, d = 30 μm, $[C]$ = 2,5 vol%; c) T_f= -90°C, d = 90μm, $[C]$ = 2vol%; d) T_f= -81°C, d=35μm, $[C]$ = 2,5 vol%.

The similar dependences were obtained for **CCH** embedded into **CFAO** (see figure 4). Enthalpy difference is ΔH_{ea}=320±40 cal mol^{-1}.The freezing-in of the conformational transitions was observed at T_f= -93±10°C.

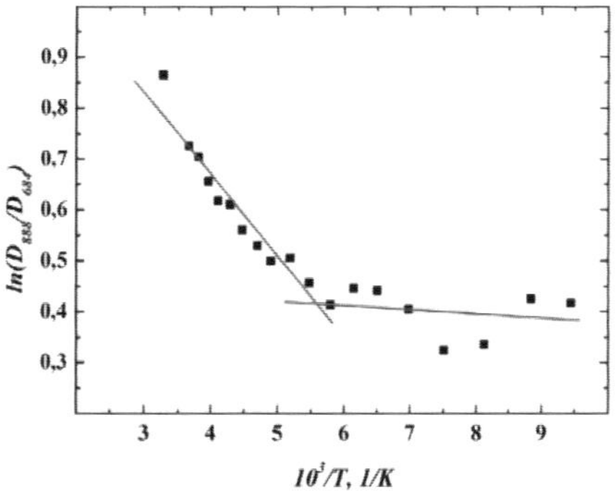

Figure 4. The dependence of $ln(D_t/D_g)$ upon T^{-1} for **CCH** embedded in **CFAO**: T_f= -93°C, d=155μm, [C] = 6 vol%.

It can be seen that the conformational mobility of **DCE** and **CCH** is freezing-in at the similar temperature, around T_f = -90°C. According to [10] β-process in **CFAO** takes place in this temperature range.

Each probe is characterized by the activation volume $V_p^{\neq}$, which is defined as a minimal free volume necessary for the internal rotation of definite groups of the probe molecule resulting in the transition from one conformation to another [5]. To estimate activation volumes for H_2Cl group in **DCE** and ClCH groups in **CCH**, we have used the method of increments [11, 12], the obtained values being equal to 24 Å^3 and 31 Å^3, respectively. The freezing of such probe molecules gives evidence to the termination of local mobility of the groups of similar and greater volume; the groups having lower volumes preserve their local mobility and freeze at lower temperatures. The smallest kinetic unit of the oligomer under study is the end NO_2.group having the activation volume 14 Å^3. Taking into account the above consideration, one may conclude that the freezing of the probe mobility is not connected with the motion of these groups. Other groups with larger activation volumes, which provide the free volume generation in the oligomer, are the benzene rings in the main and side chains. The "kinetic volume" of benzene ring (C_4H_4) is estimated to be equalto 59 Å^3. Since at T_f ~ –90°C the local mobility of the probe with smaller activation volume terminates, this gives evidence to the lack of the free volume in the oligomer. This statement is in agreement with the conclusion made in [10] on the basis of molecular simulation study, that relaxation β-process in **CFAO** is related to the freezing of the benzene rings motion.

References

[1] Dalton L R 2003 *J.Phys.: Condens. Matter.* **15** p 897
[2] Vilitis O, Muzicante I, Rutkis M and Vembris A 2012 *Latv. J. Phys. Tech. Sci.***2** p 70
[3] Vembris A, Rutkis M and Laizane E 2011 *Mol. Cryst. Liq. Cryst.* **485** p 873
[4] Shi W, Ding Y J, Fang C, Pan Q and Gu Q 2003 *Appl. Phys.* **A 77** p 567
[5] Fishman A I, Remizov A B and Stolov A A 1999 *J. Mol. Struct.* **480–481** p 303
[6] Orville-Thomas W J (Ed.) 1974 *Internal Rotation in Molecules* (New York:Wiley)
[7] Shulyndin S V, Vakhonina T A, Ivanova N V, Gubanov E F, Ustyugov A N, Fominykh O D, Estrina G A, Rozenberg B A and Zuev M B 2005 *Polym. Sci.* A **47** p 808
[8] Mizushima S, Shimanouchi T, Harada I, Abe Y and Takeuchi H 1975 *Can. J. Phys.* **53** p 2085

[9] Durig J R, Defrawy A M E, Ward R M, Guirgis G A and Gounev T K 2008 *Struct. Chem.* **19** p 579
[10] Nikonorova N A, Balakina M Yu, Fominykh O D, Pudovkin M S, Vakhonina T A, Diaz-Calleja and Vakimansky A V 2012 *Chem. Phys. Lett.* **552** p 114
[11] Askadskii AA and Kondrashchenko V I 2003 *Computational Materials Science of Polymers* (Cambridge: Int. Science Publishing)
[12] Fishman A I, Noskov A I and Stolov A A 2012 *Spectrochimica Acta Part* **A91** p 184

Pump-probe studies of absorption saturation and optical gain in Ce:LiCaAlF$_6$ ultraviolet active medium

A Galiev, V Semashko, O Akhtyamov, S Shnaidman, M Marisov, A Nizamutdinov and A Shavelev

Kazan Federal University, 18, Kremlevskaja str., 420008 Kazan, Russian Federation

E-mail: almaz.galiev.isl@gmail.com

Abstract. Aim of this work was to investigate nonlinear absorption of pump radiation in UV spectral range of Ce^{3+}:LiCaAlF$_6$ crystalline active medium. We were able to evaluate excited state absorption cross-section at pump wavelength (266 nm) and recombination probabilities by means of computer simulation.

1. Introduction

Ultraviolet (UV) spectral range appears to be fruitful in many applications as precise materials processing, analytic equipment and remote atmosphere control. But question of providing stable and reliable lasers here is still opened. From this point of view Ce^{3+} based active media experience interest because it is always prospective to achieve laser oscillation in required spectral range directly in order to manage characteristics of laser radiation by design of laser cavity [1].

Pump-probe spectroscopy appears to be one of prospective ways to investigate the potential of novel laser active media as this method allows quantitative estimation of its abilities in laser oscillation. At the same time implementing computer simulation methods to pump-probe results can give important information on dynamic processes in active media [2]. Further computer simulation of active media can give estimates of power and temporal characteristics of potential laser oscillators on the basis of investigated material.

Aim of this work was the investigation of nonlinear absorption of pump radiation and optical amplification in UV spectral range of Ce^{3+}:LiCaAlF$_6$ crystalline active medium and interpretation of experimental results on the basis of computer simulation with absorption of pump radiation from excited lasing level included in the model.

2. Experimental technique and samples

Pump-probe experiments were performed with Nd:YAG laser from Lotis TII company as a pump source. We used quadruple (266 nm) radiation to pump our Ce^{3+}:LiCaAlF$_6$ samples. Probe radiation was 289 nm lasing from Ce:LiCaAlF$_6$ laser by Ultraviolet Solutions LLC which, in order to eliminate problem of synchronizing pump and probe pulses, was pumped by the same 266 nm radiation. Delay between pump and probe pulses was controlled with use of Tektronix oscilloscope and fast detectors from Alphalas. Absorption coefficient and optical gain coefficient dependences on pump radiation flux were achieved with use of optical attenuators. Samples of Ce^{3+}:LiCaAlF$_6$ crystals were Brewster-cut plane-parallel plates 5 mm thick with two laser grade polished windows for longitudinal scheme of pump-probe experiment. Ce^{3+} ions content was 1 at.% in the melt. Samples were prepared from material grown in Kazan Federal University.

3. Pump-probe experiments

In the figure 1 an absorption saturation of Ce:LiCaAlF_6 crystal of 266 nm pump radiation is shown. Although absorption is traditionally decaying with pump flux, in our case the saturation appears at fluxes higher than it should be for interconfigurational 4f-5d transitions of Ce^{3+} ions characterized by high probabilities. Optical gain coefficient dependence on pump radiation flux for 289 nm probe radiation in Ce:LiCaAlF_6 crystal is shown in figure 1b.

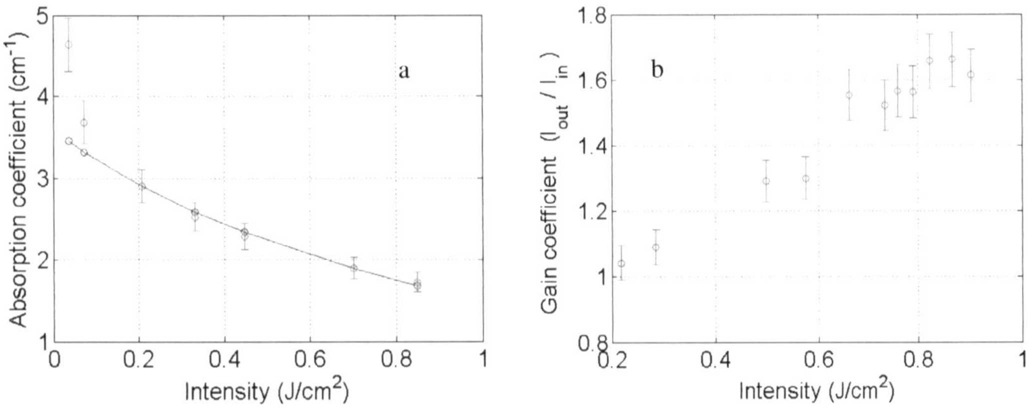

Figure 1. Absorption coefficient at 266 nm (a) and gain coefficient (b) at 289 nm dependence of Ce^{3+}:LiCaAlF_6 crystal on 266 nm pump radiation flux.

We can see that gain is increasing linearly with pump radiation flux and do not exhibit any saturation. Such dependence is characteristic for most of laser active media and evidences absence of processes of crystals optical properties degradation. Experimental errors in figure 1b are due to strong dependence of the gain coefficient on matching of pumped and probed volumes of the crystal.

4. Ce:LiCaAlF_6 active medium model and computer simulation

Ce^{3+} based active media amplify UV light due to interconfigurational 5d-4f transitions forming four-level scheme of lasing. But 266 nm photons used as a pump and ultraviolet photons of laser radiation often cause high probability of excited state absorption transition terminating in conduction band of the host lattice [3]. In order to interpret correctly the results of pump-probe experiments and to determine parameters of dynamic processes in Ce:LiCaAlF_6 active medium we have used a well known stochastic model of Ce^{3+} active medium shown in figure 2 [4,5].

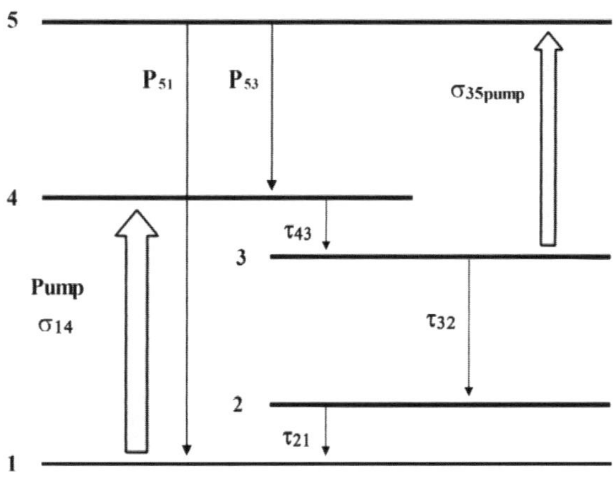

Figure 2. Stochastic model of Ce:LiCaAlF$_6$ active medium with four-level scheme of lasing based on 5d-4f transitions of Ce^{3+} ions and a level of host lattice conduction band at which excited state absorption of pump radiation transition terminate.

Our model included population inversion processes of pump radiation absorption from the lowest 4f state to excited 5d state with cross-section σ_{14} and fast thermalization at 5d state (probability $1/\tau_{43}$), processes of excited state depopulation by luminescence (probability $1/\tau_{32}$) and fast nonradiative transition to the lowest 4f state (probability $1/\tau_{21}$). Also σ_{35} is a cross-section of excited state absorption of pump radiation. Processes of relaxation of the system from conduction band are described here by probabilities P_{51} and P_{54} [5]. They correspond to recombination processes at the site of Ce^{3+} ion and leave it in the ground state and excited state, respectively. These processes make our system closed. We did not include color centers related processes in our model since we did not evidence any degradation of optical properties which could speak for accumulating of color centers absorbing incident radiation. Kinetic equations corresponding to our model are as follows:

$$\frac{d}{dt}n_1 = -Pump(t)\cdot\sigma_{14}\cdot n_1 + \frac{n_2}{\tau_{21}} + Pump(t)\cdot\sigma_{14}\cdot n_4 + P_{51}\cdot n_5 \tag{1}$$

$$\frac{d}{dt}n_2 = -\frac{n_2}{\tau_{21}} + \frac{n_3}{\tau_{32}} \tag{2}$$

$$\frac{d}{dt}n_3 = -\left[Pump(t)\cdot\sigma_{35} + \frac{1}{\tau_{32}}\right]\cdot n_3 + \frac{n_4}{\tau_{43}} + Pump(t)\cdot\sigma_{35}\cdot n_5 \tag{3}$$

$$\frac{d}{dt}n_4 = Pump(t)\cdot\sigma_{14}\cdot n_1 - \left[Pump(t)\cdot\sigma_{14} + \frac{1}{\tau_{43}}\right]\cdot n_4 + P_{54}\cdot n_5 \tag{4}$$

$$\frac{d}{dt}n_5 = Pump(t)\cdot\sigma_{35}\cdot n_3 - [Pump(t)\cdot\sigma_{35} + P_{54} + P_{51}]\cdot n_5 \tag{5}$$

Here $n_1 - n_5$ are populations at corresponding states, *Pump*(*t*) – photon flux of pump radiation given by Gaussian temporal distribution. At any moment of system evolution the sum of populations of five levels was considered to be equal to Ce^{3+} ions content N_{Ce}.

Optimization procedure consisted in search for minimal difference between experimental data on absorption and gain coefficients and data derived from computer simulation by varying parameters of dynamic processes σ_{35}, P_{51}, P_{54}. Also we varied Ce^{3+} ions content N_{Ce} as LiCaAlF$_6$ host lattice have several sites where Ce^{3+} ions could be incorporated and distribution of these sites is unknown. Parameters of spontaneous emission and gain cross-section are known from work [6], probabilities of nonradiative transitions $1/\tau_{43}$ $1/\tau_{32}$ are very high and were fixed at the level of 10^{12} s^{-1}.

Absorption coefficient computer simulation results are shown in figure 1a. Dynamic processes parameters corresponding to these results are presented in the table 1.

Table 1. Dynamic processes parameters evaluated from Ce:$LiCaAlF_6$ active medium absorption saturation by means of computer simulation

Parameter	Value
Pump radiation (266 nm) excited state absorption cross-section, cm^2	$(3{,}2\pm0{,}8)\cdot10^{-18}$
Probability of free charge carrier recombination at the Ce^{3+} site, s^{-1}	$(6{,}3\pm1{,}3)\cdot10^{8}$
Probability of free charge carrier recombination at the Ce^{3+} site terminated at excited state of Ce^{3+}, s^{-1}	$(3{,}7\pm1{,}2)\cdot10^{8}$
Ce^{3+} ions content, cm^{-3}	$(4{,}5\pm0{,}5)\cdot10^{17}$

One can see poor agreement of experimental and simulated absorption coefficients at lower pump fluxes. We explain this by the influence of pump-induced color centers which add their absorption significantly when lower pump flux do not bleach them. Excited state absorption cross-section of Ce^{3+} ions derived by us at 266 nm wavelength in $LiCaAlF_6$ has a value which is typical for Ce^{3+} doped crystals [5,6] but it is not higher than a gain cross-section, and this is proved by stable UV laser action for this active medium with 266 nm pumping. Probabilities of free charges recombination at Ce^{3+} site both have values of the same order as spontaneous 5d-4f emission of Ce^{3+} which is fast enough to make no influence on loss accumulation in the active medium but causes loss in laser efficiency when part of excitation is wasted to induce free charge carriers in conduction band and do not take part in creating the population inversion due to high rates of stimulated emission transitions. Also we were able to determine Ce^{3+} ions content for sites absorbing at 266 nm. The obtained value corresponds to low segregation coefficient of Ce^{3+} ions in $LiCaAlF_6$ crystals which leads to absorption coefficient of about 5 cm^{-1} at 266 nm for 1 at. % of Ce^{3+} in the melt.

5. Conclusion

In this work the results of nonlinear absorption and optical gain studies in Ce:$LiCaAlF_6$ active medium are presented. With the help of computer simulation of four-level scheme of Ce^{3+}-based active medium with excited state absorption included in the model we were able to evaluate parameters of excitation losses from excited 5d level of Ce^{3+} at pump wavelength and parameters of free charge carriers recombination. We have revealed that laser action efficiency is affected by excited state absorption in Ce:$LiCaAlF_6$ but remaining laser oscillation could be stable due to the absence of losses accumulation effect. Our results and computer simulation methodic could be implemented in further works on design of laser oscillators on the basis of Ce:$LiCaAlF_6$ active medium.

Acknowledgments

This work was supported by grants RFBR 12-02-31729 and Ministry of Science and Education of Russia.

References

[1] Takao O et al 2010 *Nature Photonics* **4** 767
[2] Semashko V V 2005 *Physics of the Solid State* **47** 1507
[3] Hamilton D S 1985 *Tunable solid state lasers* ed. P. Hammerling, A. B. Budger, A. Pinto (Berlin) 80
[4] Pogatshnik G J at al 1987 *Phys. Rev. B* **36** 8251
[5] Kirysheva S A et al 2011 *Proceedings SPIE* **7994** 79940G
[6] Marshall C D 1994 *J. Opt. Soc. Am. B* **11** 2054

Journal of Physics: Conference Series **478** (2013) 012025 doi:10.1088/1742-6596/478/1/012025

Studying the fidelity of quantum memory based on control-field angular scanning

L Chicherova[1] and A Kalachev[1,2]

[1]Kazan Federal University, 8, Kremlevskaya str., 420008, Kazan, Russia
[2]Zavoisky Physical-Technical Institute, 10/7, Sibirsky tract str., 420029, Kazan, Russia

E-mail: luchia131@gmail.com

Abstract. The fidelity of multimode cavity-assisted quantum storage based on control-field angular scanning is analysed as a function of an input pulse temporal shape, direction of propagation of the control field and spatial structure of the signal field. It is shown by numerical simulation that available range of the angular scanning is reduced with increasing the transverse mode index of the field to be stored. Outside of this range, the output field may contain contributions from undesirable transverse modes due to the cross-talk, which results in reducing efficiency and fidelity.

1. Introduction

In recent years much effort has been directed toward the implementation of quantum memories for light (see the reviews [1–4]). In particular, storage and retrieval of single photons is expected to be necessary for scalable linear-optical quantum computers, efficient quantum repeaters and deterministic single-photon sources. To store and recall optical pulses, such as single-photon wave packets, one can take advantage of an inhomogeneous broadened atomic transition (storage via photon echo) or can use a modulated control field amplitude which matches an input pulse (storage via electromagnetically induced transparency or off-resonant Raman interaction). Recently, another approach has been proposed [5–7] which requires neither inhomogeneous broadening nor temporal modulation of the control field amplitude, but resorts to continuous phase-matching control in an extended resonant medium. When considering off-resonant Raman interaction of a single-photon wave packet and a classical control field in a three-level atomic medium, the phase-matching control can be achieved by modulating the refractive index of the resonant medium [5] or by modulating the direction of propagation of the control field [6, 7]. In particular, a scheme for spatially multimode cavity-assisted quantum storage has been developed in [7]. The angular scanning does not require a direct control of atomic levels and may be realized in materials that do not allow linear Stark or Zeeman effects, while enclosing an atomic ensemble in a cavity makes it possible to achieve high efficiency of quantum storage with optically thin materials and realize the proposed scheme, e.g., in rare-earth-ion-doped crystals. Regarding spatially multimode storage, it is of importance for multiplexing in quantum repeaters and holographic quantum computers.

In the present work, we study the fidelity of the cavity-assisted quantum storage based on control-field angular scanning as a function of an input pulse temporal shape, direction of propagation of the control field and spatial structure of the signal field.

2. The model and basic equations

The idea of the considered quantum storage scheme is illustrated in figure 1. Let the wave vector of the control field be changed in time during the interaction of atoms with the input field to be stored. Then the wave vector of the spin coherence created via off-resonant Raman interaction on the transition $|1\rangle - |2\rangle$ also becomes a function of time so that during storage the single-photon wave packet is mapped into a superposition of (orthogonal) spin waves with different wave vectors. Retrieval is achieved by off-resonant spontaneous Raman scattering of the control field rotated within the same angle range. If direction of rotation during retrieval is the same as that during storage, the output pulse reconstructs the input one, while in the case of the reversed rotation the output pulse becomes a time-reversed replica of the input one. Such storage is possible, provided that the rate of control field angular scanning is sufficiently large, which means that the time interval between creations of orthogonal spin waves is shorter than characteristic time of an input pulse shape.

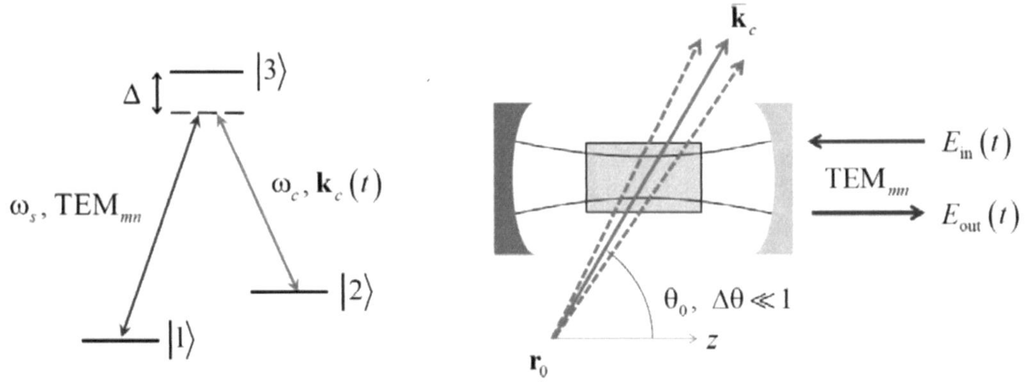

Figure 1. Energy diagram (left) and geometry (right) illustrating cavity-assisted off-resonant Raman interaction between a strong control field with a wave vector $\mathbf{k}_c(t)$, a weak quantum field in a transverse mode TEM_{mn} and a three-level atomic medium. $\mathbf{r}_0$ stands for the point where the phase shift of the control field remains zero during the rotation, $\bar{\mathbf{k}}_c$ corresponds to the average wave vector, and θ_0 is the angle between $\bar{\mathbf{k}}_c$ and the cavity axis z. Mirrors are forming a single-ended cavity for the quantum field and fully transmitting for the control field.

The multimode cavity field may be described by a set of the mode functions

$$u_{mnp}(\mathbf{r}) = e^{iq_p z} u_{mn}(\mathbf{r}), \tag{1}$$

where $q_p = 2\pi p / L$, $p \in \mathbb{Z}$, and L is the total cavity length. We restrict the consideration of the cavity field to a single longitudinal mode and suppose that different transverse modes have equal frequencies. The off-resonant interaction between the signal field and atomic system under conditions of rotated controlled field is described by the following equations [7]:

$$\begin{cases} \dfrac{\partial}{\partial t} E_{mn} = -\kappa_{mn} E_{mn} + \sqrt{2\kappa_{mn}}\, E_{mn,\text{in}}(t) + ig\sqrt{N} \sum\limits_{m',n',p} S_{m'n'p} B_{mn,m'n'p}(t), \\ \dfrac{\partial}{\partial t} S_{m'n'p} = -\gamma S_{m'n'p} + ig^* \sqrt{N} \sum\limits_{mn} E_{mn}\, B^*_{mn,m'n'p}(t), \end{cases} \tag{2}$$

where E_{mn} ($E_{mn,\mathrm{in}}$) is the slow-varying cavity (input) field amplitude, S_{mnp} is the slow-varying spin coherence amplitude, $2\kappa_{mn}$ is the cavity decay rate, γ is the spin coherence decay rate, g is the atomic-field coupling constant, and N is the number of atoms. The indexes m,n,p correspond to the mode functions $u_{mnp}(\mathbf{r})$. The coupling between different field modes and spin modes is described by the coefficients

$$B_{mn,m'n'p}(t)=\frac{1}{N}\int d\mathbf{r}\, e^{i\phi(\mathbf{r},t)}e^{iq_p z}u^*_{mn}(\mathbf{r})u_{m'n'}(\mathbf{r}), \tag{3}$$

where integration is performed over the volume of the atomic system, and $\phi(\mathbf{r},t)$ is a phase shift due to the rotation of the control field. The atomic ensemble is supposed to fill the cavity and to have a length $L_z = L$ along the cavity axis z.

In what follows, we assume that the control field propagates in the (x,z) plane at some angle θ_0 to the signal field, its wave vector is rotated within a small angle $\Delta\theta$ during the time interval T, which is the duration of the storage and retrieval process, and the rotation center (phase stationary point) $\mathbf{r}_0$ is located at the center of the sample, where the coordinate system originates. In this case, $\bar{\mathbf{k}}_c = (k_c \sin\theta_0, 0, k_c\cos\theta_0)$ and $\phi(\mathbf{r},t)=k_c(x\cos\theta_0 - z\sin\theta_0)(\Delta\theta/T)t$. Finally, we consider Hermite-Gaussian modes and assume that the origin of the z axis coincides with the beam waist (see [7] for details).

3. Fidelity of quantum storage

We are interested in the fidelity of quantum storage, which is given by

$$F=\eta F', \tag{4}$$

where

$$F'=\frac{\left|\sum_{mn}\int_0^{+\infty}dt\left\langle E^\dagger_{mn,\mathrm{out}}(t)E_{mn,\mathrm{in}}(\pm t)\right\rangle\right|^2}{N_{\mathrm{in}}N_{\mathrm{out}}} \tag{5}$$

is the correlation between the input and output pulse envelopes (the minus sign corresponds to retrieval with time reversal),

$$N_{\mathrm{in}}=\sum_{mn}\int_{-\infty}^{0}dt\left\langle E^\dagger_{mn,\mathrm{in}}(t)E_{mn,\mathrm{in}}(t)\right\rangle,$$
$$N_{\mathrm{out}}=\sum_{mn}\int_{0}^{+\infty}dt\left\langle E^\dagger_{mn,\mathrm{out}}(t)E_{mn,\mathrm{out}}(t)\right\rangle, \tag{6}$$

and $\eta = N_{\mathrm{out}}/N_{\mathrm{in}}$ is the total storage efficiency. It is assumed that the storage process terminates at the moment $t=0$, while the retrieval process begins at this moment of time. The input-output relation is $E_{mn,\mathrm{out}}(t)=\sqrt{2\kappa_{mn}}E_{mn}(t)-E_{mn,in}(t)$.

Let us start with the fidelity as a function of an input pulse temporal shape by considering storage and retrieval in a specific transverse mode. According to Eqs. (2), at different moments of time the signal field effectively interacts with spin waves having different longitudinal components of wave vectors q_p. The time interval between creations of orthogonal spin waves is equal to [7]

$$\delta=\frac{T}{\Delta\theta}\frac{\lambda_c}{L_z\sin\theta_0}, \tag{7}$$

where λ_c is the wavelength of the control field. Numerical simulations show that a Gaussian pulse with a duration as short as δ can be stored and recalled with a close to unity efficiency [5, 7]. Figure 2 illustrates the fidelity of quantum storage as a function of the pulse duration δ_p (FWHM) for different pulse shapes: Gaussian pulse $E_{\text{in}}(t) \propto \exp\left(-(t-t_0)^2 \sigma^2 / 2\right)$ of duration $\delta_p \approx 1.67/\sigma$, hyperbolic secant pulse $E_{\text{in}}(t) \propto \text{sech}\left((t-t_0)\sigma\right)$ of duration $\delta_p \approx 1.76/\sigma$, and double-exponential pulse $E_{\text{in}}(t) \propto \exp\left(-|t-t_0|\sigma\right)$ of duration $\delta_p \approx 1.39/\sigma$. Since the reversal switching time δ is equivalent to the inhomogeneous linewidth of the atomic transition, the fidelity reduces when the pulse duration becomes smaller than the switching time. Pulses with more narrow spectral distribution give higher fidelity at the same value of duration. In the case of the bad cavity limit, $2\kappa\delta \gg 1$, the optimal short pulse proves to be double exponential one, which matches the Lorentzian absorption profile.

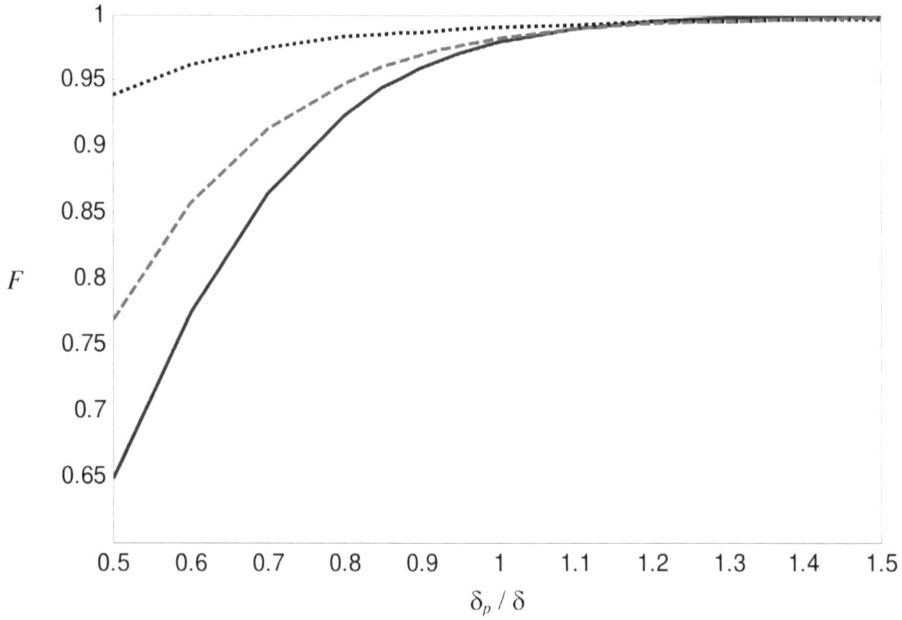

Figure 2. The fidelity of quantum storage as a function of the pulse duration δ_p (FWHM) for different pulse shapes: Gaussian pulse (blue solid line), hyperbolic secant pulse (red dashed line) and double-exponential pulse (black dotted line). The plots are obtained by numerically solving Eqs. (2) for a single TEM_{mn} mode with the following values of parameters $\theta_0 = \pi/2$, $T = 30\delta$, $\gamma = 0$ and $\kappa\delta = 5$.

In a general case, the input field may excite different transverse modes. The spatially multimode storage and retrieval of single photons is possible provided that [7]

$$|\tan\theta_0| \gg \frac{\pi w_0}{L_z}\frac{T}{\delta}\sqrt{m+1}, \tag{8}$$

where w_0 is the waist size, and m is the largest transverse mode index of the field to be stored. Under the condition (8), different transverse modes interact with the atomic system independently and can be stored and recalled in parallel. To satisfy Eq. (8), the angle θ_0 should be large enough, which means that a transverse control field proves to be the ideal one. These features are illustrated in figure 3, which presents the fidelity of quantum storage as a function of the angle between the control field and signal field θ_0 for different transverse modes TEM_{m0}. The input pulse is assumed to be a Gaussian one of duration $\delta_p = \delta$. The fidelity decreases when the angle θ_0 approaches the minimum value corresponding to the right hand side of Eq. (8). The larger the mode index m, the smaller available range of the angular scanning. Outside of this range, the output field may contain contributions from undesirable transverse modes due to the cross-talk, which results in reducing efficiency and fidelity.

Figure 4 illustrates storage and retrieval of a single Gaussian pulse in a third order transverse mode TEM_{30}. In the case of a transverse control field, fidelity and efficiency of quantum storage are both close to unity ($\eta = 0.99$, $F = 0.99$). However, they are significantly reduced when the angle is small, e.g., $\eta = 0.63$ and $F = 0.43$ for $\theta_0 = 10^\circ$. In the latter case, the output field can be expressed as a superposition of different transverse modes (71% of TEM_{30}, 13% of TEM_{40}, 8% of TEM_{20}, etc.), which reduces correlation between the input and output pulse shapes.

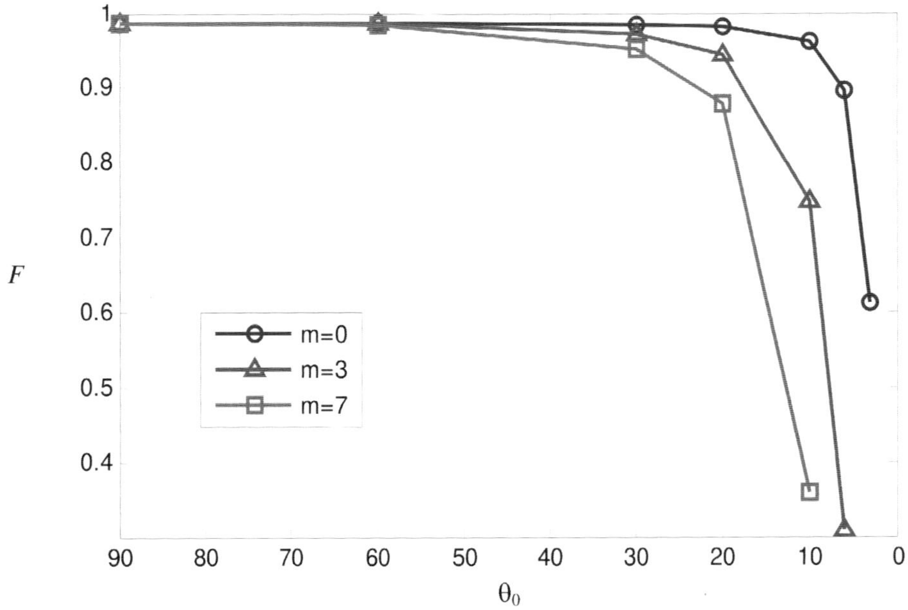

Figure 3. The fidelity of quantum storage as a function of the angle between the control field and signal field for different transverse modes TEM_{m0}. The plot is obtained as a result of numerically solving Eqs. (2) with the following values of parameters: $T = 3\delta$, $\delta_p = \delta$, $\pi w_0 / L_z = 0.1$, $\gamma = 0$ and $\kappa_{mn}\delta = 5$.

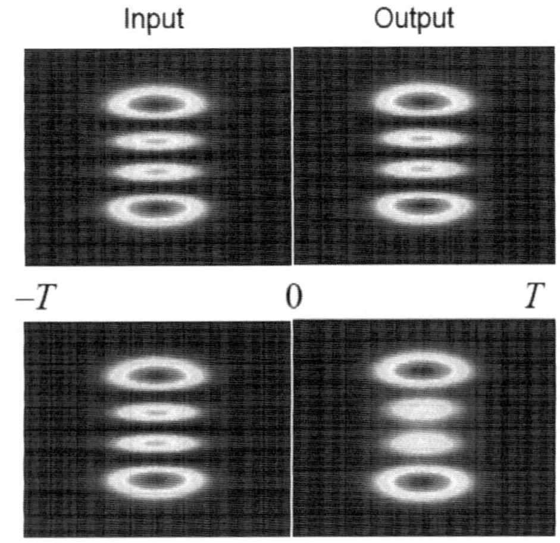

Figure 4. Illustrating storage and retrieval of a single Gaussian pulse in a third order transverse mode TEM_{30}. The left (right) column corresponds to the input (output) pulse. The first line presents storage and retrieval in the case of a transverse control field, $\theta_0 = 90^\circ$, the second line – in the case of $\theta_0 = 10^\circ$. The horizontal axis represents time, while the vertical axis represents the transverse coordinate x. The plots are obtained by numerically solving Eqs. (2) with the conditions $T = 10\delta$, $\delta_p = 2\delta$, $\pi w_0 / L_z = 0.1$, $\gamma = 0$ and $\kappa_{mn}\delta = 5$.

4. Conclusion

The fidelity of multimode cavity-assisted quantum storage based on control-field angular scanning, which has been proposed recently in [7], is analysed numerically as a function of an input pulse temporal shape, direction of propagation of the control field and spatial structure of the input field. It is shown that the larger the transverse mode index of the field to be stored, the smaller available range of the angular scanning. Regarding temporal pulse shapes, under conditions of the bad cavity limit double-exponential pulses provide higher fidelity for small values of pulse duration.

Acknowledgments

This research was supported by the RFBR (Grant No. 12-02-00651) and the Program of the Presidium of RAS 'Quantum mesoscopic and disordered structures'.

References

[1] Lvovsky A I, Sanders B C and Tittel W 2009 *Nat. Photon.* **3** 706
[2] Hammerer K, Sørensen A S and Polzik E S 2010 *Rev. Mod. Phys.* **82** 1041
[3] Tittel W, Afzelius M, Cone R L, Chanelière T, Kröll S, Moiseev S A and Sellars M 2010 *Laser Photon. Rev.* **4** 244
[4] Simon C, Afzelius M, Appel J, Boyer de la Giroday A, Dewhurst S J, Gisin N, Hu C Y, Jelezko F, Kröll S, Müller J H, Nunn J, Polzik E S, Rarity J G, De Riedmatten H, Rosenfeld W, Shields A J, Sköld N, Stevenson R M, Thew R, Walmsley I A, Weber M C, Weinfurter H, Wrachtrup J, and Young R J 2010 *Eur. Phys. J.* D **58** 1
[5] Kalachev A and Kocharovskaya O 2011 *Phys. Rev.* A **83** 053849
[6] Zhang X, Kalachev A and Kocharovskaya O 2013 *Phys. Rev.* A **87** 013811
[7] Kalachev A and Kocharovskaya O 2013 *Phys. Rev.* A **88** 033846

Journal of Physics: Conference Series **478** (2013) 012026 doi:10.1088/1742-6596/478/1/012026

Magnetic and magnetoelastic properties of $LiDyF_4$ single crystals

I V Romanova[1,2], S L Korableva[1], V I Krotov[3], B Z Malkin[1], I R Mukhamedshin[1,2], H Suzuki[2], M S Tagirov[1]

[1]Kazan Federal University, Kazan, 420008, Kremlevskaya 18, Russia
[2]Kanazawa University, Kanazawa, 920-11, Kakuma-machi, Japan
[3]Kazan State Power Engineering University, Kazan, 420066, Krasnoselskaya, 51, Russia

E-mail: Irina.Choustova@kpfu.ru

Abstract. Temperature and magnetic field dependences of the magnetization of $LiDyF_4$ single crystal were measured with a dc-SQUID magnetometer. The longitudinal magnetostriction was measured by the capacitance-bridge method. A giant compression of the crystal lattice up to 0.1% was observed in the magnetic field of 1 T along the [110] direction at 4.2 K. Experimental data are well reproduced by simulations based on the microscopic models of the single-ion magnetoelastic and inter-ion multipole interactions.

1. Introduction

Double lithium-rare earth fluorides which crystallize in the tetragonal scheelite structure attract a lot of interest as model objects in physics of dipolar magnets [1]. The unit cell of $LiLnF_4$ contains two magnetically equivalent trivalent lanthanide ions at sites with the S_4 point symmetry. Single crystal $LiDyF_4$ is a dipolar XY-antiferromagnet with Dy^{3+} magnetic moments normal to the crystal symmetry axis (T_N=0.62 K) [1]. It was shown recently that $LiDyF_4$ is a promising material for Faraday rotators in the ultraviolet-visible wavelength region [2]. The analysis of magnetic dipolar and exchange interactions between the Dy^{3+} ions and of the magnetic structure was presented in [3] based on the measurements of the low temperature specific heat. The energy levels crossover effects in strong pulse magnetic fields on the magnetization of $LiDyF_4$ were studied in [4].

The main goal of the present study was to elucidate the role of magnetoelastic and multipole interactions in formation of the magnetization, magnetostriction and the energy level structure of Dy^{3+} ions in $LiDyF_4$ crystals in the external magnetic fields.

2. Experimental details

Single crystals of $LiDyF_4$ were grown by Bridgeman-Stockbarger method. The samples were oriented by means of X-ray diffractometer. The sample used for magnetization measurements was shaped by polishing as a sphere to acquire a definite demagnetizing factor. It was fixed in Stycast 1266 epoxy (the accuracy of orientation was ±3°) in order to prevent rotation in strong magnetic fields. Temperature and field dependences of the single crystal magnetization in the temperature range of 2 - 300 K and the magnetic fields in the range of 0 - 5 T applied along and perpendicular to the c-axis were measured using a dc-SQUID magnetometer MPSM-2 (Quantum Design).

The measured temperature and field dependences of the magnetization in $LiDyF_4$ single-crystals are compared with the results of simulations in figures 1 and 2.

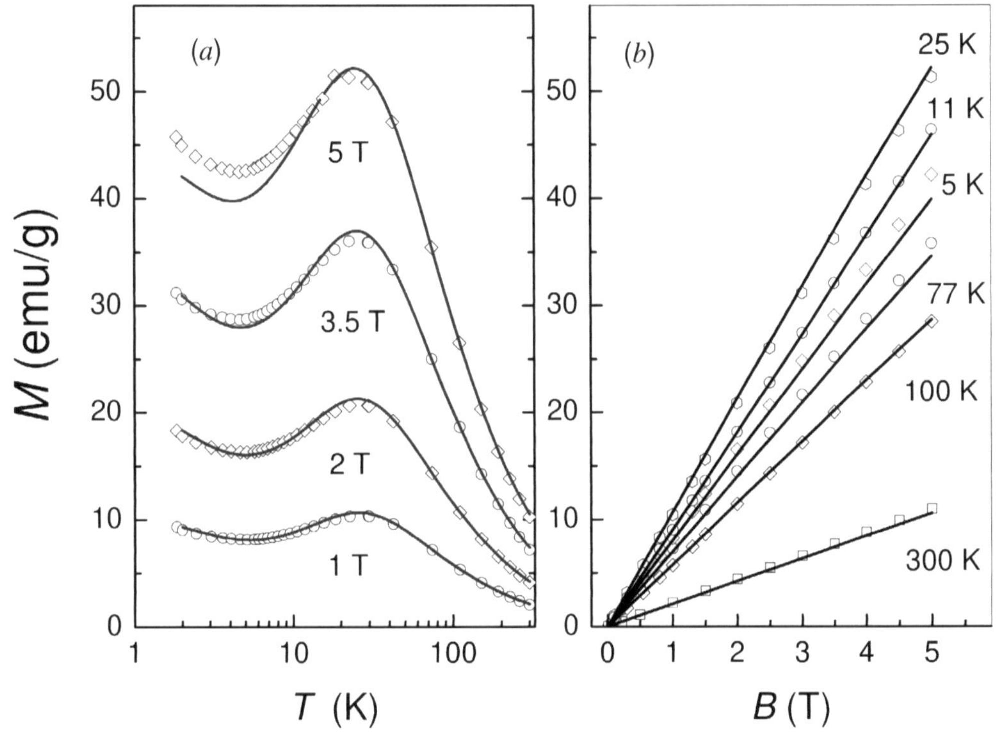

Figure 1. Measured (symbols) and calculated (solid curves) temperature (*a*) and magnetic field (*b*) dependences of the magnetization in $LiDyF_4$ for different values of magnetic fields parallel to the crystal c-axis and temperatures.

The longitudinal magnetostriction was measured at the temperature 4.2 K as a function of the external magnetic field varied from zero up to 1 T. The measurements were performed by the capacitance-bridge method [5]. The crystal was placed in a helium cryostat at the center of a superconducting solenoid and was connected to the measuring device by means of a quartz rod. The signals induced by the change of the sample length with increase of the magnetic field were directly registered with an X-Y recorder; a voltage proportional to the current through the solenoid was applied to the X input. The sign of the deformation was determined and the sensitivity of the apparatus was calibrated by means of control measurements of the magnetostriction of the polycrystalline nickel. To obtain the measurement results independent on the displacement of the sample under the action of the torque due to the interaction of the anisotropic magnetic moment of the Dy^{3+} ions with the magnetic field, the sample was secured with vacuum grease or silicone oil. The experiments were performed on cylindrical samples of 5-6 mm diameter and 6-10 mm length, with finely polished flat and parallel surfaces. The axes of the samples were oriented along the crystallographic directions [100] or [110] by the X-ray diffraction with an accuracy not worse than 3°. The results of the measurements are presented in figure 3.

3. Experimental results and discussion: magnetization and magnetostriction in $LiDyF_4$

The measured temperature and magnetic field dependences were analyzed by making use of the effective single-ion Hamiltonian

$$H_{eff} = H^{(0)} + H^{(P)} \tag{1}$$

where

$$H^{(0)} = H_0 + H_{cf} + H_Z, \tag{2}$$

$$H^{(P)} = \{\sum_{\alpha\beta} B_{eff,\alpha\beta} e_{\alpha\beta} - <O>_0 : [\frac{n}{v}\sum_{rr'} D_{eff,\alpha}(r)\hat{a}^{-1}_{\alpha\beta}(r,r')D_{eff,\beta}(r') + \lambda : (1+q:\lambda)^{-1}]\} : O \tag{3}$$

operating in the space of states of the $4f^9$ electronic configuration of the Dy^{3+} ion. Here H_0 is the free ion standard Hamiltonian defined by the parameters of electrostastic, spin-orbit and interconfigurational interactions presented in [6]. The crystal field Hamiltonian in the crystallographic system of coordinates is determined by the set of seven crystal field parameters B_p^k,

$$H_{cf} = B_2^0 O_2^0 + B_4^0 O_4^0 + B_4^4 O_4^4 + B_4^{-4} O_4^{-4} + B_6^0 O_6^0 + B_6^4 O_6^4 + B_6^{-4} O_6^{-4}, \tag{4}$$

where O_p^k are linear combinations of spherical tensor operators [7]. The third term in (2) is the electronic Zeeman energy in the external magnetic field $\boldsymbol{B}$, $H_Z = \mu_B \mathbf{B}(\mathbf{L} + 2\mathbf{S})$, here μ_B is the Bohr magneton, $\boldsymbol{L}$ and $\boldsymbol{S}$ are orbital and spin moments of an ion, respectively.

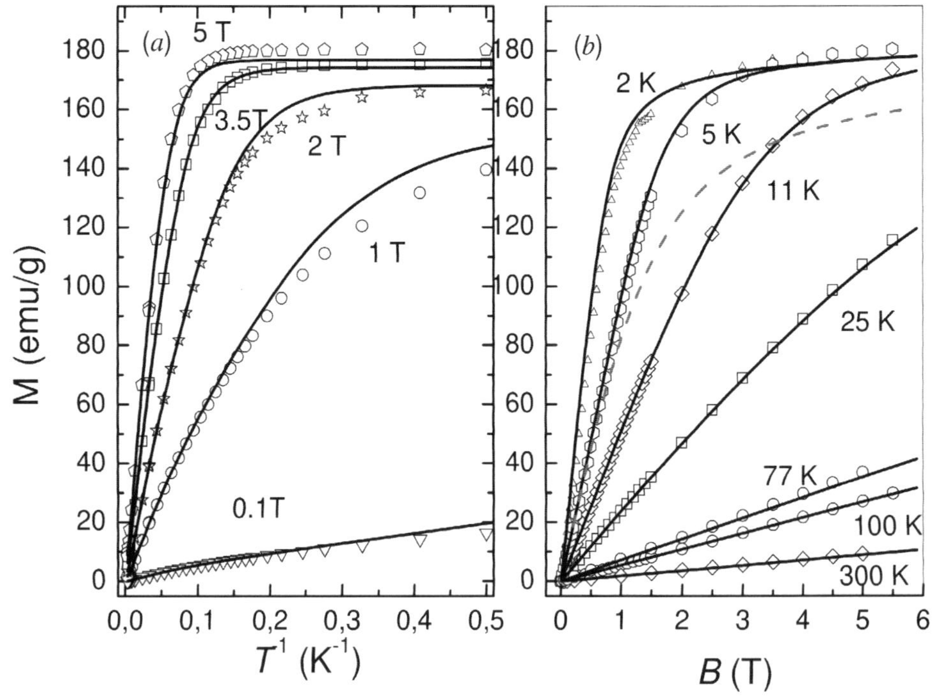

Figure 2. Measured (symbols) and calculated (solid curves) temperature (*a*) and magnetic field (*b*) dependences of the magnetization in $LiDyF_4$ for the magnetic fields parallel to the crystal *a*-axis. The dashed curve represents the calculated magnetization at *T*=5 K when the electron-deformation interaction is not taken into account.

The Hamiltonian (3) contains terms corresponding to multipole interactions between the Dy^{3+} ions and the electron-deformation interaction. Here a colon symbol means a convolution over the indices of the spherical tensors ($\sum_{pk} A_p^k O_p^k = A:O$), n=2 is the number of the magnetically equivalent paramagnetic ions in the unit cell with a volume v. The matrix q is the generalized multipole susceptibility depending on temperature T:

$$q_{pp'}^{kk'} = \frac{1}{k_B T}[<O_p^k>_0 <O_{p'}^{k'}>_0 - \sum_i \langle\psi_i|\rho O_p^k|\psi_i\rangle\langle\psi_i|O_{p'}^{k'}|\psi_i\rangle] \\ + \sum_{i,j\neq i}(\varepsilon_i-\varepsilon_j)^{-1}[\langle\psi_i|\rho O_p^k|\psi_j\rangle\langle\psi_j|O_{p'}^{k'}|\psi_i\rangle + \langle\psi_i|\rho O_{p'}^{k'}|\psi_j\rangle\langle\psi_j|O_p^k|\psi_i\rangle], \tag{5}$$

here ε_i and ψ_i are eigenvalues and eigenfunctions of the Hamiltonian $H^{(0)}$, k_B is the Boltzmann constant, a symbol $<...>_0$ means quantum statistical averaging with the equilibrium density matrix $\rho = \exp(-H^{(0)}/k_B T)/Tr(\exp(-H^{(0)}/k_B T))$. The electron-deformation interaction linear in components of the deformation tensor $e_{\alpha\beta}$ and the sublattice displacements $w_\alpha(r)$ $(r = 1 \div 11)$ is defined by the parameters $B_{p,\alpha\beta}^k$ and $D_{p,\alpha}^k(r)$:

$$H_{el-def} = \sum_{pk}[\sum_{\alpha\beta} B_{p,\alpha\beta}^k e_{\alpha\beta} + \sum_{\alpha,r} D_{p,\alpha}^k(r) w_\alpha(r)] O_p^k. \tag{6}$$

These parameters are renormalized due to coupling between the macro- and microdeformations determined by the tensors $b_{\alpha\beta,\gamma}(r)$ [8, 9] and by the multipole interactions. In particular, we use in (3) the effective parameters of the electron-deformation interaction

$$D_{eff\ p,\alpha}^k(r) = \sum_{p'k'}[(1+\lambda:q)^{-1}]_{pp'}^{kk'} D_{p',\alpha}^{k'}(r), \tag{7}$$

$$B_{eff\ p,\alpha\beta}^k = \sum_{p'k'}[(1+\lambda:q)^{-1}]_{pp'}^{kk'} \hat{B}_{p',\alpha\beta}^{k'}, \tag{8}$$

where

$$\hat{B}_{p,\alpha\beta}^k = B_{p,\alpha\beta}^k - \sum_{rr'\gamma\delta} b_{\alpha\beta,\gamma}(r) a_{\gamma\delta}^{-1}(r,r') D_{p,\delta}^k(r'), \tag{9}$$

and $a_{\alpha\beta}(r,r')$ is the dynamic matrix of the crystal lattice at the Brillouin zone centre.

The multipole interactions were considered in the framework of the self-consistent field approximation neglecting second-order terms in fluctuations $O_p^k(Ls) - <O_p^k(Ls)>$ of the operators $O_p^k(Ls)$ relating to ions at sites s in unit cells L. The corresponding average values $<O_p^k(Ls)> = <O_p^k>$ were eliminated from the explicit expression for the free energy F of the crystal by making use of the equilibrium conditions $\partial F/\partial <O_p^k> = 0$. We assume that the multipole interactions are induced by the phonon exchange between the paramagnetic ions. In this case components of the matrix λ introduced in (3) can be written as follows [10]:

$$\lambda_{pp'}^{kk'} = \frac{1}{n}\sum_{ss'}\left[\sum_{j_o}\frac{B_p^k(s,0j_o)B_{p'}^{k'}(s',0j_o)}{\omega_{\mathbf{0}j_o}^2} - \frac{\delta_{ss'}}{N}\sum_{\mathbf{q}j}\frac{B_p^k(s,\mathbf{q}j)B_{p'}^{k'}(s',-\mathbf{q}j)}{\omega_{\mathbf{q}j}^2}\right], \tag{10}$$

here the sum in the first term is taken over optical branches of the lattice vibrational spectrum, N is the number of the cells, $\omega_{\mathbf{q}j}$ is the frequency of a phonon with the wave vector $\boldsymbol{q}$, $B_p^k(s,\mathbf{q}j)$ are parameters of the electron-phonon interaction linear in lattice normal coordinates.

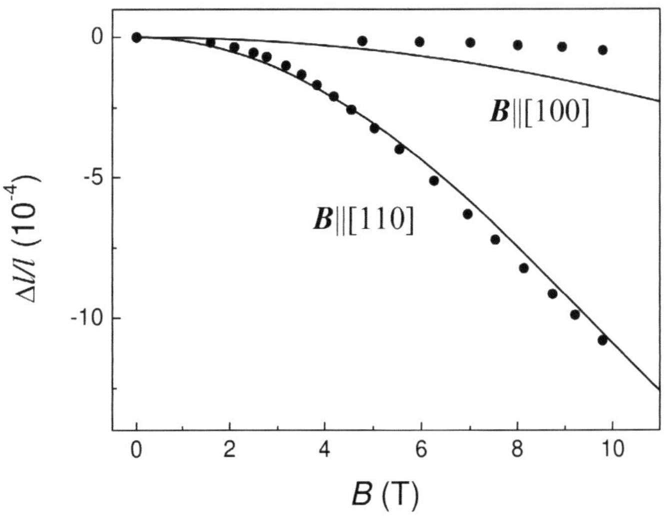

Figure 3. Measured (symbols) and calculated (solid curves) longitudinal magnetostriction of $LiDyF_4$ in the external magnetic fields along the [100] and [110] directions, T=4.2 K.

The free energy of the elastically deformed single crystal (per unit cell) written in second order in the perturbation (3) has the following form:

$$F = \frac{v}{2}\mathbf{eCe} + nF_0 - \frac{n}{2} < O >_0 : [\lambda : (1 + q : \lambda)^{-1} + \frac{n}{v}\sum_{rr'} \hat{\mathbf{D}}(r)\hat{\mathbf{a}}^{-1}(r,r')\hat{\mathbf{D}}(r')] : < O >_0 + n < O >_0 : \mathbf{B}_{eff}\mathbf{e} \quad (11)$$

where F_0 is the free energy of a single ion with the Hamiltonian $H^{(0)}$ and

$$C_{\alpha\beta\gamma\delta}(T,\mathbf{B}) = C^{(0)}_{\alpha\beta\gamma\delta} + \frac{n}{v}\hat{B}_{\alpha\beta} : q : (1 + \lambda : q)^{-1} : \hat{B}_{\gamma\delta} \quad (12)$$

are renormalized elastic constants.

The calculations of the average magnetic moments

$$\mathbf{M} = \mu_B Tr[(\mathbf{L} + 2\mathbf{S})\exp(-H_{eff} / k_B T)] / Tr[\exp(-H_{eff} / k_B T)] \quad (13)$$

in the present work were carried out by making use of the numerical diagonalization of the Hamiltonian (1) in the total space of 2002 electronic states of the $4f^9$ configuration. The magnetic dipolar interactions between the Dy^{3+} ions were accounted for by introducing the local magnetic fields

$$\mathbf{B}_{loc}(s) = \mathbf{B} + \sum_{s'}[\mathbf{Q}(s,s') - \frac{4\pi}{3v}N_d]\mathbf{M}(s'), \quad (14)$$

where $\mathbf{M}(s)$ is the magnetic moment of an ion belonging to the s-sublattice, N_d is the demagnetizing factor and $\mathbf{Q}(s,s')$ are the corresponding lattice sums.

To analyze the magnetostriction, we calculated the relative changes of the crystal size $\Delta l / l = \sum_{\alpha\beta} n_\alpha n_\beta e_{\alpha\beta}$ along a unit vector with directional cosines n_α where variations of components of the deformation tensor with the magnetic field were determined from the minimum condition of the free energy (11):

$$\mathbf{e}(\mathbf{B}) = -\frac{n}{v}[\mathbf{SB}_{eff} :< O >_0|_{\mathbf{B}} - \mathbf{SB}_{eff} :< O >_0|_{\mathbf{B}=0}] \tag{15}$$

(here $\mathbf{S} = \mathbf{C}^{-1}$ is the compliance tensor).

Table 1. Parameters of the interaction between the Dy^{3+} ions and lattice deformations of B_g symmetry in $LiDyF_4$ (cm^{-1}). The effective parameters are given for zero magnetic field and T=4.2 K.

	p k							
	2 2	2 -2	4 2	4 -2	6 2	6 -2	6 6	6 -6
$\hat{B}_p^k(B_g^1)$	1997	2355	-887	1200	-135	-427	-731	-691
$B_{eff\,p}^{\;k}(B_g^1)$	1617	3817	-1497	1766	-322	-549	-614	-1022
$\hat{B}_p^k(B_g^2)$	4114	-1144	-1120	1652	-528	-375	-841	-878
$B_{eff\,p}^{\;k}(B_g^2)$	4614	-4406	337	168	-126	-4	-997	-98

The calculations are essentially simplified when making use of symmetry properties of a system. Indeed, we worked with linear combinations of the deformation tensor $e(A_g^1) = e_{zz}$, $e(A_g^2) = (e_{xx} + e_{yy})/2$, $e(B_g^1) = e_{xx} - e_{yy}$, $e(B_g^2) = e_{xy}$, $e_1(E_g) = e_{xz}$, $e_2(E_g) = e_{yz}$ and the sublattice displacements corresponding to irreducible representations A_g, B_g, E_g of the lattice factor group C_{4h}. In particular, the magnetic field directed along the c-axis induces only totally symmetric A_g deformations, while A_g and rhombic B_g deformations are induced by fields perpendicular to the c-axis.

Table 2. Parameters of the mutipole interactions $\lambda_{ij} = \lambda_{ji}$ (cm^{-1}).

$i \backslash j$	1	2	3	4	5	6	7	8
1	-72.02	-54.76	48.49	-54.84	10.18	26.5	22.5	29.96
2		62.98	-17.76	10.88	-7.127	3.516	16.07	-6.565
3			-2.0	4.572	-3.149	-0.482	-13.64	-4.882
4				-28.56	-1.775	4.771	14.09	1.137
5					-1.935	-1.158	-4.147	-1.7
6						-1.932	-5.529	-6.8
7							-6.556	-9.053
8								-6.287

As follows from the previous studies of magnetoelastic effects in double lithium-rare earth fluorides [1, 9], interactions of rare earth ions with the lattice deformations of the B_g symmetry play the dominant role. The corresponding terms in the Hamiltonian (6) involve only eight operators $O_2^2{=}O_1$, $O_2^{-2}{=}O_2$, $O_4^2{=}O_3$, $O_4^{-2}{=}O_4$, $O_6^2{=}O_5$, $O_6^{-2}{=}O_6$, $O_6^6{=}O_7$, $O_6^{-6}{=}O_8$. The corresponding parameters of the multipole interaction, elements of the matrix λ, in accordance with (10) where we neglect dependences of the phonon frequencies in the optical branches on the wave vectors, can be written as

$$\lambda_{ij} = \sum_{\upsilon=1:5} \frac{B_i(B_g^\upsilon)B_j(B_g^\upsilon)}{\omega(B_g^\upsilon)^2} - \sum_{\upsilon=1:4} \frac{B_i(A_u^\upsilon)B_j(A_u^\upsilon)}{\omega(A_u^\upsilon)^2} - \frac{1}{N}\sum_{\mathbf{q}j_a} \frac{B_i(1,\mathbf{q}j_a)B_j(1,-\mathbf{q}j_a)}{\omega_{\mathbf{q}j_a}^2}. \tag{16}$$

The parameters (16) were estimated using frequencies of even and odd optical vibrations of B_g and A_u symmetry at the Brillouin zone centre of $LiLnF_4$ crystals and the corresponding coupling constants

$B_i(B_g^{\upsilon})$, $B_i(A_u^{\upsilon})$ which were determined earlier from the analysis of temperature and magnetic field effects in Raman scattering spectra [11] and calculated in the framework of the exchange charge model [8]. The summation over acoustic vibrations in the last term in (16) was completed by making use of the long-wave approximation that gave a possibility to connect the electron-phonon interaction constants $B_i(1,\mathbf{q}j_a)$ with the parameters of the electron-deformation interaction (9) which were determined from piezospectroscopic measurements in [12, 13] (see table 1). The obtained values of the parameters λ_{ij} are presented in table 2.

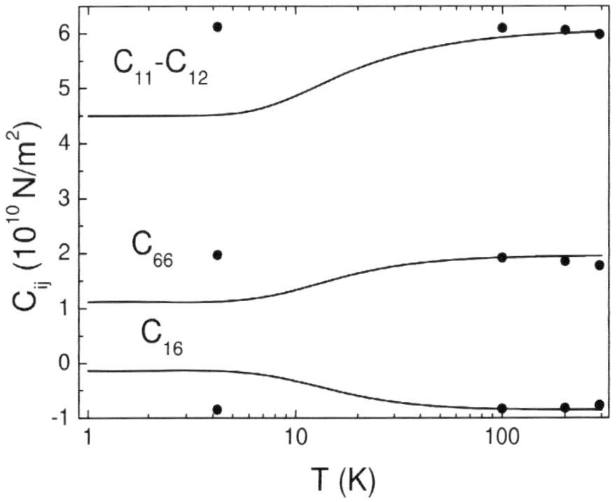

Figure 4. Calculated temperature dependences of the elastic constants in zero magnetic field. The values of the corresponding elastic constants of $LiYF_4$ are represented by symbols [15].

Relatively small corrections of the crystal field parameters obtained earlier from the analysis of the optical spectra in dilute isomorphic crystals $LiYF_4$:Dy [14] and transformed by the proper rotation of the coordinate frame around the *c*-axis [8] gave us a possibility to fit satisfactorily the measured temperature and field dependences of the magnetization in the magnetic fields parallel to the *c*-axis (see table 3). The observed not monotone behavior of the magnetization at low temperatures with the maximum close to 25 K and the minimum at 5 K (see figure 1*a*) is caused by the specific relation between the corresponding components of the *g*-tensors of the ground ($g_{\|}$=1.15) and the first excited ($g_{\|}$=5.3) crystal field sublevels of the $^6H_{15/2}$ multiplet of the Dy^{3+} ions with a gap of 24 K and the redistribution of populations of these Kramers doublets with temperature variations.

Table 3. Crystal field parameters B_p^k (cm^{-1}) for $LiDyF_4$.

p	*k*	Ref. [14]	This work
2	0	165	160
4	0	-88	-83
6	0	-4.4	-4.6
4	4	-980	-751
4	-4	0	-641
6	4	-427	-360
6	-4	-65	-243

The essential role of the electron-deformation interaction in the formation of the magnetization of the concentrated rare-earth paramagnets at low temperatures becomes apparent when analyzing the magnetization of $LiDyF_4$ induced by the magnetic fields perpendicular to the *c*-axis (figure 2). In this case, as it is seen in figure 2*b*, when accounting only for the Zeeman interaction, we obtain strongly underestimated magnetic moments of the Dy^{3+} ions in the fields exceeding 1 T.

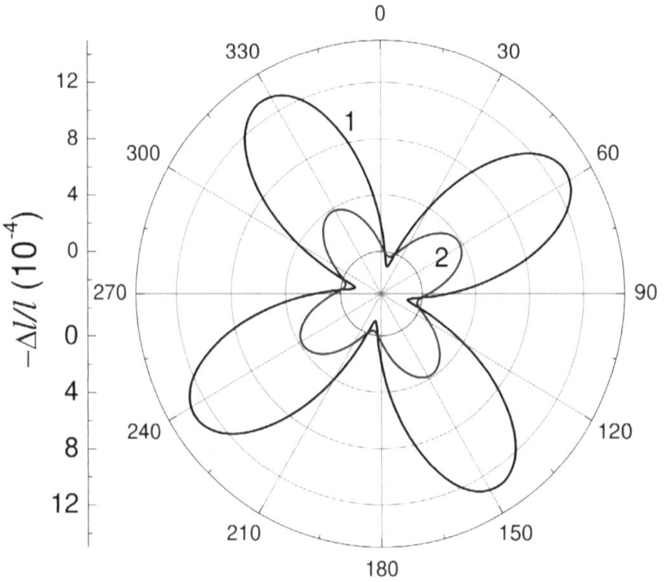

Figure 5. Calculated angular dependences of the magnetostriction in the magnetic fields B=1 T (curve 1) and 0.5 T (curve 2), $\mathbf{B}\perp[001]$, T=4.2 K.

It should be noted that the multipole interactions bring the remarkable renormalization of the electron-deformation interaction (see table 1), and one can expect a strong softening of the $LiDyF_4$ crystal lattice similar to that observed in $DyVO_4$ [10]. The calculated temperature dependences of the elastic constants of $LiDyF_4$ which determine lattice strains induced by deformations of B_g symmetry are presented in figure 4. Indeed, the elastic constants change remarkably when the temperature decreases down to 10 K but remain almost constant at lower temperatures.

The calculated relative changes of the crystal dimensions along the magnetic field oriented in the basis plane of the crystal lattice and declined from the *a*-axis by the angle φ

$$\Delta l / l = e(A_g^2) + \frac{1}{2}\cos(2\varphi)e(B_g^1) + \sin(2\varphi)e(B_g^2) \tag{16}$$

are presented in figure 5. The calculated magnetostriction is strongly anisotropic and reaches giant values of an order of 10^{-3} in the magnetic field field ~1 T at T=4.2 K. The calculated field dependence of the longitudinal magnetostriction agrees well with the experimental data (see figure 3).

4. Conclusion

The temperature and field dependences of the magnetization and field dependences of the strongly anisotropic magnetostriction in $LiDyF_4$ single crystals were measured. The giant magnetostriction of 10^{-3} was observed in rather small magnetic field of 1 Tesla directed along the [110] crystallographic axis at liquid helium temperature. The sets of parameters which define interactions of the Dy^{3+} ions with the crystal field and the crystal lattice uniform deformations as well as the parameters of the multipole interactions between the Dy^{3+} ions were determined. The results of measurements are well reproduced by simulations. The magnetoelastic interaction in $LiDyF_4$ contributes essentially to the magnetization in external magnetic fields oriented in the basis plane of the crystal lattice at low temperatures.

Acknowledgements
The authors are grateful to V.A. Shustov for the X-Ray orientation of the samples. This work was partially supported by the RFBR grant №12-02-00372-a.

References

[1] Aminov L K, Malkin B Z and Teplov M A 1996 Magnetic properties of nonmetallic lanthanide compounds *Handbook on the Physics and Chemistry of the Rare-Earths.* **22** Amsterdam: North-Holland 295
[2] Vasyliev V, Villora E G, Nakamura M, Sugahara Y and Shimamura K 2012 *Optics Express* **20** 14460
[3] Mennenga G, de Jongh L J, Huiskamp W J and Laursen I 1984 *J. Mag. Mag. Mat.* **44** 48
[4] Kazei Z A, Snegirev V V, Chanieva R I et al. 2006 *Sol. St. Phys.* **48** 682
[5] Sokolov V I 1967 *Pribory i Tekhnika Eksperimenta (Instruments and Experimental Techniques)* **2** 181
[6] Crosswhite H M, Crosswhite H 1984 *J. Opt. Soc. Am. B* **1** 246
[7] Klekovkina V V, Zakirov A R, Malkin B Z, Kasatkina L A 2011 *J. Phys.: Conf. Ser.* **324** 012036
[8] Bumagina L A, Krotov V I, Malkin B Z, Khasanov A Kh 1981 *Sov. Phys. JETP* **54** 792
[9] Al'tshuler S A, Malkin B Z, Teplov M A, Terpilovskii D N 1985 Magnetoelastic interactions in rare-earth paramagnets $LiLnF_4$ *Soviet Scientific Reviews, Section A, Physics Reviews* **6** New York: Harwood Academic Publishers 159
[10] Elliott R J, Harley R T, Hayes W, Smith S R P 1972 *Proc. Roy. Soc. Lond. A.* **328** 217
[11] Kupchikov A K, Malkin B Z, Natadze A L, Ryskin A I 1987 *Phys. Sol. State* **29** 3335
[12] Vinokurov A V, Malkin B Z, Pominov A I, Stolov A L 1986 *Phys. Sol. State* **28** 381
[13] Bertaina S, Barbara B, Giraud R et al. 2006 *Phys. Rev. B.* **74** 184421
[14] Davidova M P, Zhdanovich S B, Kazakov B N, Korableva S L, Stolov A L 1977 *Optica i spektr.* **42** 577
[15] Blanchfield P, Saunders G A 1979 *J. Phys. C: Solid State Phys.* **12** 4673

AUTHOR INDEX

Akhmadeev, A. A. 67
Akhtyamov, O. 122
Anikushina, T. A. 24
Balakina, M. Yu 117
Barantsev, K. A. 45
Basharov, A. M. 59
Biktagirov, T. B. 6
Bizyaev, D. A 112
Bukharaev, A. A. 112
Bunkov, Yury 15
Chelyshev, Yu A. 6
Chicherova, L. 126
Ellseeva, Svetlana V. 49
Eremchev, I. Yu 29
Fedyanin, V. V. 39
Fishman, A. I. 117
Gafurov, M. R. 1, 6
Gainutdinov, R. Kh 87, 92, 97
Galiev, A. 122
Gazizov, A. R. 77
Gilmutdinov, A. Kh 71
Goovaerts, E. 1
Ivanin, K. V. 104
Izotov, V. V. 11
Kalachev, A. 126
Karimullin, K. R. 39
Kazantseva, E. V. 63
Khamadeev, M. A. 87, 97
Khanipov, T. F. 112
Kharintsev, S. S. 1, 77, 82
Klimusheva, G. V. 39
Korableva, S. L. 132
Koryukin, A. V. 67
Krotov, V. I. 132
Leontyev, A. V. 104
Litvinov, A. N. 45
Lobkov, V. S. 104
Magarian, K. A. 39
Maimistov, A. I. 63
Malkin, B. Z. 132
Mamin, G. V. 1, 6, 11
Marisov, M. 122
Mitrofanova, T. G. 104
Morozov, M. V. 71
Mukhamedshinl, I R 132
Mutygullina, A. A. 92, 97
Naumov, A. V. 24
Nefedyeva, K. L. 117
Nikiforov, V. G. 108
Nurgazizov, N. I. 112
Orlinskii, S. B. 1, 6, 11
Osad'Ko, I. S. 29
Osin, Yu N. 1, 6
Ostatochnikov, Vladimir A. 49
Petrova, A. S. 97
Rogov, A. M. 77
Romanova, I. V. 132
Sa, G. M. 108
Salakhov, M. Kh 1, 6, 67, 71, 77, 82, 87
Samartsev, V. V. 104
Savochkin, I. V 55
Semashko, V 122
Sementsov, Dmitrij I. 49
Shchukina, A. L. 29
Shmelev, A. G. 108
Shnaidman, S. 122
Sukhorukov, A. P. 55
Suzuki, H. 132
Tagirov, M. 15, 132
Vakhonina, T. A. 117
Vasilchenko, V. E. 82
Vasilieva, I. A. 39
Volodin, M. A 11
Yavkin, B. V 1

Institute of Physics
Dirac House, Temple Back
Bristol BS1 6BE UK

proceedings.com